Geoinformation

W9-AXR-076

Surveying and mapping have recently undergone a transition from discipline-oriented technologies, such as geodesy, surveying, photogrammetry and cartography, into the methodology-oriented integrated discipline of geoinformatics. This is based on GPS positioning, remote sensing, digital photography for data acquisition and GIS for data manipulation and data output. Gottfried Konecny attempts to present the required basic background for remote sensing, digital photogrammetry and GIS in the new geoinformatics concept in which the different methodologies must be combined.

For remote sensing, the basic fundamentals are the properties of electro-magnetic radiation and their interaction with matter. This radiation is received by sensors and platforms in analogue or digital form, and is subject to image processing. In photogrammetry, the stereo-concept is used for the location of information in 3D. With the advent of high-resolution satellite systems in stereo the theory of analytical photogrammetry restituting 2D image information into 3D is of increasing importance, merging the remote sensing approach with that of photogrammetry. The result of the restitution is a direct input into geographic information systems in vector or in raster form. The fundamentals of these are described in detail, with an emphasis on global, regional and local applications. In the context of data integration, a short introduction to the GPS satellite positioning system is provided.

This textbook will appeal to a wide range of readers from advanced undergraduates to all professionals in the growing field of geoinformation.

Gottfried Konecny is Professor at the University of Hannover in Germany and former President of the International Society of Photogrammetry and Remote Sensing.

Geoinformation

Remote sensing, photogrammetry
and geographic information systems

Gottfried Konecny

London and New York

First published 2003
by Taylor & Francis
11 New Fetter Lane, London EC4P 4EE

Simultaneously published in the USA and Canada
by Taylor & Francis Inc
29 West 35th Street, New York, NY 10001

Taylor & Francis is an imprint of the Taylor & Francis Group

© 2003 Gottfried Konecny

Typeset in Sabon by Wearset Ltd, Boldon, Tyne and Wear
Printed and bound in Great Britain by TJ International Ltd, Padstow,
Cornwall

British Library Cataloguing in Publication Data
A catalogue record for this book is available from the British Library

Library of Congress Cataloging in Publication Data
A catalog record for this book has been requested

ISBN 0–415–23795–5 (pbk)
ISBN 0–415–23794–7 (hbk)

Contents

Illustrations

Figures

Plates

Tables

Acknowledgements

I would like to acknowledge the help I received from a number of individuals.

Ms Gesine Boettcher, secretary of the Institute for Photogrammetry and GeoInformation (IPI) at the University of Hannover, who typed the manuscript.

Ms Susan Penzold, cartography student in Berlin, who drew the line artwork using *Corel Draw*.

My successor in office at the University of Hannover, Prof. Dr. Christian Heipke and his 'right hand', Dr. Karsten Jacobsen, who supported me within the Institute's infrastructure.

Dr. Peter Lohmann and Dr. Jacobsen of the Institute, who have proofread the manuscript and have helped in the compilation of reference literature.

Prof. Dr. Manfred Schroeder of DLR, Oberpfaffenhofen, who has proofread the remote sensing portion of the book.

Ms Sarah Kramer of the publishers, Taylor & Francis, who helped with the obtaining of permissions for the use of illustrations in the book.

On a personal level, I would like to thank my wife Lieselotte, who has fully supported me in my activities.

Gottfried Konecny,
Hannover, November 2001

1   Introduction

Surveying and mapping has recently undergone a transition from discipline oriented technologies, such as geodesy, surveying, photogrammetry and cartography into a methodology oriented integrated discipline of geoinformation based on GPS positioning, remote sensing, digital photography for data acquisition and GIS for data manipulation and data output. This book attempts to present the required basic background for remote sensing, digital photogrammetry and geographic information systems in the new geoinformation concept, in which the different methodologies must be combined depending on efficiency and cost to provide spatial information required for sustainable development. In some countries this concept is referred to as 'geomatics'.

For remote sensing the basic fundamentals are the properties of electro-magnetic radiation and their interaction with matter. This radiation is received by sensors on platforms in analogue or digital form to result in images, which are subject to image processing. In photogrammetry the stereo-concept is used for the location of the information in three dimensions. With the advent of high resolution satellite systems in stereo, the theory of analytical photogrammetry, restituting two-dimensional image information into three dimensions, is of increasing importance merging the remote sensing approach with that of photogrammetry.

The result of the restitution is a direct input into geographic information systems in vector or in raster form. The application of these is possible at the global, regional and local levels.

Data integration is made possible by geocoding, in which the GPS satellite positioning system plays an increasing role. Cost considerations allow a judgement on which of the alternate technologies can lead to an efficient provision of the required data.

Surveying and mapping in transition to geoinformation

Geodesy

Geodesy, according to F.R. Helmert (1880), is the science of measurement and mapping of the earth's surface. This involves, first, the determination of a reference surface onto which details of mapping can be fixed.

In ancient Greece (Homer, 800 BC) the earth's surface was believed to be a disk surrounded by the oceans. But, not long after, Pythagoras (550 BC) and Aristotle (350 BC) postulated that the earth was a sphere. The first attempt to measure the dimensions of a spherical earth was made by Eratosthenes, a Greek resident of Alexandria around 200 BC. At Syene (today's Assuan) located at the Tropic of Cancer at a latitude of 23.5° the sun reflected from a deep well on June 21, while it would not do so in Alexandria at a latitude of 31.1°. Erathosthenes measured the distance between the two cities along the meridian by cart-wheel computing the earth's spherical radius as 5909 km. Meridional arcs were later also measured in China (AD 725) and in the caliphate of Baghdad (AD 827).

Until the Renaissance, Christianity insisted on a geocentric concept, and the determination of the earth's shape was not considered important. In the Netherlands Willebrord Snellius resumed the ancient ideas about measuring the dimensions of a spherical earth using a meridional arc, which he measured by the new concept of triangulation, in which long distances were derived by trigonometry from angular measurements in triangles. The scale was derived from one accurately surveyed small triangle side, which was measured as a base by tape.

The astronomers of the Renaissance (Copernicus (1500), Kepler (1600) and Galilei (1600)) along with the gravitational theories of Newton (1700) postulated that the earth's figure must be an ellipsoid, and that its flattening could be determined by two meridional arcs at high and low latitude. While the first verification in France (1683–1718) failed due to measurement errors, the measurement of meridional arcs in Lapland and Peru (1736–1737) verified an ellipsoidal shape of the earth. Distances on the ellipsoid could consequently be determined by the astronomical observations of latitude and longitude at the respective points on the earth's surface.

Laplace (1802), C.F. Gauss (1828) and F.W. Bessel (1837), however, recognized that astronomic observations were influenced by the local gravity field due to mass irregularities of the earth's crust. This was confirmed among others by G. Everest observing huge deflections of the vertical in the Himalayas. This led to the establishment of local best-fitting reference ellipsoids for positional surveys of individual countries.

In the simplest case latitude and longitude was astronomically observed at a fundamental point, and an astronomical azimuth was measured to a second point in the triangulation network spanning a country. Within the triangulation network, at least one side was measured by distance measuring devices on the ground. For the determination of a best-fitting ellipsoid, several astronomic observation stations and several base lines were used. The coordinates of all triangulated and monumented points were calculated and least squares adjusted on the reference ellipsoid with chosen dimensions, e.g. for half axis major a and for half axis major b or the flattening

$$f = \frac{a - b}{a}:$$

Clarke 1880 $a = 6\,378\,249\,\text{m}, b = 6\,356\,515\,\text{m}$
Bessel $a = 6\,377\,879\,\text{m}, f = 1/298.61$
Hayford $a = 6\,378\,388\,\text{m}, f = 1/297$
Krassovskij $a = 6\,378\,295\,\text{m}, f = 1/298.4$

On the chosen reference ellipsoid, the ellipsoidal latitudes and longitudes were obtained for all points of the first order triangulation network. This network was subsequently densified to second, third and fourth order by lower order triangulation.

The survey accuracy of these triangulation networks of first to fourth order was relatively high, depending on the observational practices, but discrepancies between best fitting ellipsoids of neighbouring countries were in the order of tens of metres.

For the purpose of mapping, the ellipsoidal coordinates were projected into projection coordinates. Due to the nature of mapping in which local angular distortions cannot be tolerated, conformal projections are chosen:

- for circular countries (the Netherlands, the Province of New Brunswick in Canada), the stereographic projection.
- for N–S elongated countries, the 3° Transverse Mercator projection tangent to a meridian, every 3 degrees. Due to its first use by C.F. Gauss and its practical introduction by Krüger, the projection is called the Gauss–Krüger projection. It is applied for meridians 3° apart in longitude in several strips. The projection found wide use for the mapping of Germany, South Africa and many countries worldwide.
- for E–W elongated countries (France) the Lambert Conic Conformal Projection was applied to several parallels.
- the Lambert conic conformal projection may also be obliquely applied, e.g. in Switzerland.
- for worldwide mapping, mainly for military mapping requirements, the UTM projection (a 6° Transverse Mercator Projection) is applied. The formulation is the same as for the Gauss–Krüger projection, with the exception that the principal meridian (here every 6 degrees) has a scale factor of 0.9996 rather than 1 used for the Gauss-Krüger projection.

Since the earth's gravity field influences the flow of water on the earth's surface, ellipsoidal coordinates without appropriate reductions cannot be used for practical height determinations. Height reference systems are therefore separate from position reference systems based on reference ellipsoids.

An ideal reference surface would be the equipotential surface of the resting oceans, called 'the geoid'. Due to earth tides influenced by the moon and planets, due to ocean currents and winds influenced by climate and meteorology, this surface is never resting. For this reason the various countries engaged in mapping systems have created their own vertical reference systems by observing mean sea level tides at tidal benchmarks. Spirit levelling extended the elevations in level loops of first order over the mapping area of a country to monumented benchmarks. These level loop observations, corrected by at least normal gravity, could be densified by lower order levelling to second, third and fourth order. As is the case for positions, differences of several metres in height values may be the result of the different height reference systems of different countries.

The different reference systems for position and height still used for mapping in the countries of the world are in transition, changing into a new reference frame of three- or four-dimensional geodesy. This has become possible through the introduction of the US Navy Navstar Global Positioning Systems (GPS) in the 1980s. It now consists of twenty-four orbiting satellites at an altitude of 20 200 km. These orbit at an inclination of 55° for 12 hours, allowing a view, in a direct line of sight, of at least four of these satellites from any observation point on the earth's surface for 24 hours of the day.

Each of the satellites transmits timed signals on two carrier waves with 19.05 cm and 24.45 cm wavelengths. The carrier waves are modulated with codes containing the particular satellite's ephemeris data with its corrections. The US Defense Department has access to the precise P-code suitable for real-time military operations. Civilian users can utilize the less precise C/A code carried by the 19.05 cm carrier wave.

When three satellites with known orbital positions transmit signals to a ground receiver, the observed distances permit an intersection of 3D-coordinates on the earth's surface. Since the satellite clocks are not synchronized, an additional space distance from a fourth satellite is required for 3D positioning.

The calculations are based on an earth mass centred reference ellipsoid determined by an observation network by the US Department of Defense, the WGS 84 with the following dimensions:

$$a = 6 378 137 \text{ m}$$
$$f = 1/298.257 223 563$$

Local reference ellipsoids used in the various countries differ in coordinate positions by several 100 m with WGS 84 coordinates.

P-code observations may be used in real time to accuracies in the dm range. C/A codes are capable of determining positions at the 5 m level unless the satellite clock signals are artificially disturbed by the military satellite system operators, as was the case during the 1990 to 2000 period. This disturbance was called the 'selective availability (SSA)'. It deteriorated the C/A code signals to 100 m accuracies in position and to 150 m in height.

To overcome this lack of dependability, more elaborate receivers were developed in the civilian market, which observed the phases of the carrier waves, using the C/A codes only to obtain approximate spatial distances and to eliminate ambiguities when using the phase measurements. The principle of measurement at a mobile rover station thus became that of relative positioning with respect to a permanently operating master reference station.

In static mode (observing over longer duration periods) positional accuracies in the range of several millimetres could be achieved for distances closer than 10 km. For long distances over several hundreds of kilometres, accuracies in the 1 cm to 2 cm range could be obtained by the simultaneous observation of networks.

Relative observations in networks are able to minimize ionospheric and tropospheric transmission effects. Satellite clock errors may be eliminated using double differences.

Multiple effects may be eliminated by the careful choice of observation points. This has encouraged the international civilian community to establish an International Terrestrial Reference Frame (ITRF) of over 500 permanently observing GPS stations worldwide. The absolute position of ITRF is combined with the observation of an International Celestial Reference Frame (ICRF), in which the absolute orientation of ITRF is controlled by stellar observations using radio astronomy (quasars, VLBI).

The existence of an ITRF gives the opportunity to monitor changes of plate tectonic movements of the earth's crust. Thus ITRF is determined at a specified epoch (e.g. ITRF 1993, ITRF 1997, ITRF 2000, etc.), in which local plate deformations can be observed which exceed centimetre accuracies.

The existence of ITRF has encouraged mapping agencies throughout the world to establish new continental control networks and to densify them into national reference systems.

In Europe, thirty-six ITRF stations were selected in 1989 to create the European Reference Frame ETRF 89. This reference frame served to re-observe national networks with differential GPS, such as the DREF 91 in Germany, which permitted the setting up of a network of permanently observing GPS reception stations SAPOS, with points about 50 km apart.

Networking solutions, such as those offered by the companies Geo++ or Terrasat permit the use of transmitted corrections to rover stations observing GPS-phase signals in real time kinematic mode. These enable positioning in ETRF to 1 cm accuracy in latitude and longitude and to 2 cm accuracy in height at any point of the country, where the GPS signals may be observed. Austria, Sweden (SWEPOS) and the Emirate of Dubai have introduced similar systems. Therefore GPS has become a new tool for detailed local surveys and its updating.

Thus the problem of providing control for local surveys and mapping operations has been reduced to making coordinate conversions from local reference systems to new geodetic frameworks for data, which have previously been collected.

A detailed coverage of the modern geodetic concept with the mathematical tools has been given in the book *Geodesy* by Wolfgang Torge. A treatise of satellite geodesy is contained in the book *Satellite Geodesy* by Günter Seeber.

Surveying

While geodesy's main goal was to determine the size and shape of the earth, and to provide control for the orientation of subsequent surveys on the earth's surface, it was the aim of survey technology to provide tools for the positioning of detailed objects.

The first direct distance measurements have been in place since the Babylonians and the Romans. Direct distance observations by tapes and chains, as they have been used in former centuries, have made way to the preference of angular measurement by theodolites, which permitted the use of trigonometry to calculate distances in overdetermined angular triangulation networks. Even for detailed terrestrial topographic surveys, instruments were developed in the early 1900s, the so-called tacheometers, which were able to measure distances rapidly in polar mode by optical means, combining these with directional measurements.

In the 1950s, the direct measurement of distances became possible by the invention of electronic distance-measuring devices. The tellurometer, developed in South Africa, utilized microwaves as a carrier onto which measurement phases of different wavelengths were modulated. The Swedish geodimeter used light as a carrier and later most manufacturers used infrared carrier waves. These distance-measuring capabilities were soon combined with directional measurement devices, known from theodolites in the form of electronic tacheometers. The automatic coding of directional and distance measurements was perfected in the so-called 'total stations'.

Nowadays it is possible to integrate GPS receivers and PC graphic capabilities in a single instrument, the so-called 'Electronic field book', which can be used effectively for updating vector graphics or for orienting oneself in the field with the help of digitized and displayed images. For the survey of heights, simple route survey devices, such as the barometer, are considered to be too inaccurate according to today's standards.

Levelling is still the prime source of accurate height data, even though spirit level devices have gradually been replaced by instruments with automatic compensators assuring a horizontal line of sight by the force of gravity. The optically observed level rods read by the operator have like-wise been automated by digitally coded reading devices to permit more rapid levelling operations.

Nevertheless, ground surveys without the use of GPS are still considered very expensive, and they are thus only suitable for the detailed survey of relatively small areas.

Remote sensing

Remote sensing can be considered as the identification or survey of objects by indirect means using naturally existing or artificially created force fields. Of most significant impact are systems using force fields of the electro-magnetic spectrum which permit the user to directionally separate the reflected energy from the object in images.

The first sensor capable of storing an image, which could be later inter-preted, was the photographic emulsion, discovered by Nièpce and Daguerre in 1839. When images were projected through lenses onto the photographic emulsion, the photographic camera became the first practical remote sensing device around 1850.

As early as 1859, photographs taken from balloons were used for mili-tary applications in the battle of Solferino in Italy and later during the American Civil War. Only after the invention of the aircraft in 1903 by the Wright brothers did a suitable platform for aerial reconnaissance become of standard use. This was demonstrated in World War I, during which the first aerial survey camera was developed by C. Messter of the Carl Zeiss Company in Germany in 1915. Aerial photographic interpretation was extended into many application fields (glaciology, forestry, agriculture, archaeology), but during World War II it again became the primary recon-naissance tool on all sides.

In Britain and Germany, development of infrared sensing devices began, and Britain was successful in developing the first radar, in the form of the 'plan position indicator' (PPI).

Further developments were led by the United States in the post-war years, developing colour-infrared film in the 1950s. Other developments went on in side-looking airborne radar (SLAR, SAR).

In the 1960s, remote sensing efforts made use of the first satellite plat-forms. 'Tiros' was the first meteorological satellite. The lunar landing preparations for the Apollo missions of NASA had a strong remote sensing component, from sensor development through to analysis. When the lunar landing was accomplished, NASA turned its interest toward remote sensing of the earth's surface.

In 1972, the Earth Resources Technology Satellite (ERTS-1), which later was called Landsat 1, became the first remote sensing satellite with a world coverage at 80 m pixels in four spectral visible and near infrared channels.

In subsequent years both spatial and spectral resolution of the first Landsat were improved: Landsat 3, launched in 1982, had six visible and near infrared channels at 30 m pixels and one thermal channel at 120 m pixels.

Higher spatial resolution was achieved by the French Spot satellites launched since 1986 with panchromatic pixel sizes of 10 m and multispectral resolution at 20 m pixels. The Indian satellites IRS 1C and 1D reached panchromatic pixel sizes of 6 m in 1996. Even higher resolution with photographic cameras was reached from space in the US military Corona programme of the 1960s (3 m resolution) and the Russian camera KVR 1000 (2 m resolution) in 1991. Since 1999, the US commercial satellite Ikonos 2 is in orbit, which produces digital panchromatic images with 1 m pixels on the ground. This was surpassed by the US commercial satellite Quickbird, with 0.6 m pixels on the ground.

In 1978, the first coherent radar satellite Seasat was launched by the United States, but it only had a very short lifetime. In 1991, the European Space Agency (ESA) commenced a radar satellite programme ERS 1 and 2, which was followed by the Japanese radar sensor on JERS 1 in 1994, the Canadian Radarsat in 1995 and the Russian Almaz in 1995. The Space Shuttle Radar Topography mission SRTM in the year 2000 carried an American C band radar sensor and a German X band radar sensor by which large portions of the land mass of the earth were imaged at pixel sizes of 25 m to 30 m. Coherent radars are not only of interest due to their all-weather, day and night capability, but due to the possibility of deriving interferograms from adjacent or subsequent images, which permit the derivation of relative elevations at 12 m to 6 m accuracy.

A new development is the construction of hyperspectral sensing devices, which, at lower resolutions, permit the scanning of the earth in more than 1000 narrow spectral bands to identify objects by their spectral signatures.

Multisensoral, multispectral, multitemporal, and in the case of radar, even multipolarization images permit vast possibilities for image analysis in remote sensing, which are not yet fully explored.

Photogrammetry

Photogrammetry is the technology to derive geometric information of objects with the help of image measurements. While the use of perspective geometry dates back to the Italian Renaissance (Brunneleschi (1420), Piero della Francesa (1470), Leonardo da Vinci (1481) and Dürer (1525)), it needed the invention of photography to develop a usable tool. In 1859, the military photographer Aime Laussedat used terrestrial photographs for the construction of plans for the city of Paris. In 1858, the architect Meydenbauer used terrestrial photographs for the derivation of plans for the cathedral of Wetzlar in Germany, and Sebastian Finsterwalder of Munich used photographs for the survey of glaciers in the Tyrolean Alps in 1889.

Terrestrial and balloon photographs were, however, not suitable for providing a systematic coverage of the earth's surface. This was made possible by the use of aircraft combined with the use of aerial survey cameras during World War I.

Even though Sebastian Finsterwalder made a rigid reconstruction of a pair of balloon photographs of the area of Gars am Inn in 1899, using mathematical calculations of many points measured in the photos, this analytical approach, without the availability of computers, was too time consuming. Therefore, analogue instrumentation with stereo measurement was developed, which permitted fast optical or mechanical reconstruction of the photographic rays determining an object point. In the 1920s, German, Italian, French and Swiss designers developed a great variety of photogrammetric plotting instruments. These were used during World War II to meet the rapidly increased mapping demands of the nations involved in the war.

After World War II, modifications and simplifications of the optical and mechanical plotting instruments developed in Switzerland, in Britain, in France, in Italy, in the USA and the USSR played an important role in meeting basic mapping demands all over the globe. In order to diminish control requirements for mapping, aerial triangulation was developed as a method to interpolate a scarce control distribution down to the requirements of a single stereo model.

The introduction of the computer into photogrammetry in the late 1950s not only permitted the part-automation of the tasks of aerial triangulation and stereo restitution, but also aided in the increased accuracy and reliability of the restitution process.

Computer developments in the 1970s and 1980s, with increased speed and storage, finally permitted users to treat aerial images in digital form after they have been scanned by raster scanners. Digital photogrammetry combined with image processing techniques finally became the new tool to partially or fully automate point measurement, coordinate transformation, image matching for the derivation of the third dimension and for differential image rectification to create orthoimages with the geometry corresponding to a map. This technology is not only suitable to be applied to aerial photographs, but it can likewise also be used, with slight modifications, for terrestrial images or for digital images of satellite scanners.

The development of laser scanners for use in aircraft is relatively new. Successful systems have been developed to measure the distance between scanner and the terrain with accuracies in the range of 10 cm. The advantage of the technique is that it can be used in urban areas and in forests obtaining multi-levels of terrain information (terrain versus roof tops of buildings, terrain versus tree tops). A proper restitution technology, however, requires that the sensor position and the sensor orientation is accurately known.

Attempts to measure sensor position and orientation date back to the 1920s, when instrumentation for the measurement of some orientation components was made possible by devices such as the statoscope, in which barometric pressure changes were used to derive elevation differences between exposure stations. Other devices, such as the horizon camera, aimed at determining the sensor orientation. Today, in-flight GPS permits the establishment of sensor positions in real time with 15 cm accuracy. Inertial measuring units permit the determination of sensor inclination with a correspondingly high accuracy. The aim is to solve the control densification problem without the need for aerial triangulation. A historical review of the developments in photogrammetry is given in Konecny (1996).

Geographic information systems (GIS)

Geographic information systems arose from activities in four different fields:

- cartography, which attempted to automate the manually dependent map-making process by substituting the drawing work by vector digitization.
- computer graphics, which had many applications of digital vector data apart from cartography, particularly in the design of buildings, machines and facilities.
- databases, which created a general mathematical structure according to which the problems of computer graphics and computer cartography could be handled.
- remote sensing, which created immense amounts of digital image data in need of geocoded rectification and analysis (see Figure 1.1).

The first theoretical and exploratory attempts for a GIS design started in the 1960s, namely by:

- R.F. Tomlinson, who in 1968 created the first Canada Geographic Information System for an agricultural agency (ARDA).
- the Experimental Cartography Unit under Bickmore in the UK, which has attempted to automate cartography since 1963.
- the Harvard Laboratory for Computer Graphics, which has laid the theoretical foundations for successful industrial GIS developments since 1964, through their creation of Symap (e.g. Jack Dangermond for ESRI and D.F. Sinton for Intergraph).
- the Swedish attempts to establish a Land Information System for the district of Upsala.

These first personal initiatives were followed by the takeover of these ideas by governmental administrations, e.g. by:

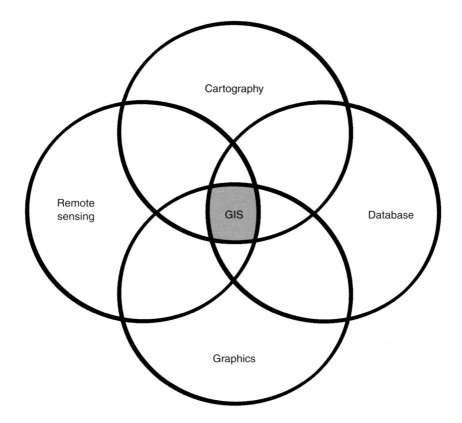

Figure 1.1 Interrelationship between GIS disciplines.

- the US Bureau of Census, from 1967, utilizing DIME files.
- the US Geological Survey, creating Digital Line Graphs.
- the attempts of the German states and their surveys and mapping coordination body, ADV, creating the ATKIS concept.

The successful introduction of the technology without subsequent strong industrial development would not have been possible. The leaders in the field from 1969 were:

- ESRI under Jack Dangermond, in the USA.
- Intergraph under Jim Meadlock, in the USA.
- Siemens in Germany.

As the developments in computing speed and computer storage grew rapidly in the 1980s and 1990s, it became obvious that an integration of

industrial efforts for user requirements was mandatory. This led to the creation of the Open GIS Consortium (OGIS) which, along with the standards organizations ISO and CEN, led to the transferability of data created
on different systems, to the integration of vector and raster data, to the
establishment of metadata bases and to the transmission of data in the
Internet or Intranet.

It was soon realized that data created and analysed for project purposes
only was a duplication of effort, and that the tendency went into the creation of base data, which had to be continuously updated not only to
provide an inventory of geospatial information, but to permit its cross-
referenced analysis and to integrate the effort into a management system
for sustainable development. A review of the historical GIS developments
is given in the book, *Geographical Information Systems*, by Longley,
Goodchild, Maguire and Rhind.

The current status of mapping in the world

The UN Secretariat has tried to monitor existing base map data for different countries and continents at different scale ranges (see Table 1.1).

A near global coverage only exists at the scale range of 1:200 000 or
1:250 000. At 1:50 000, about two-thirds of the land area is covered and at
1:25 000 about one-third. The coverage at these scales is in analogue form,
but is subject to progress in vectorization or at least in raster scanning. No
data surveys exist at large scales.

At the UN Cartographic Conference in Bangkok in 1990, a survey on
the annual update rates of these maps was presented, as shown in Table
1.2.

The conclusion of these surveys is that the update rate for the 1:50 000
map is only 2.3 per cent. This means that the average existing 1:50 000
map of the countries of the world is about 44 years old, and that the
average existing 1:25 000 map is 20 years old.

While ground survey methods applied in the nineteenth century have
been able to create national coverages in Europe within a century, aerial

Table 1.1 Status of world mapping (1990)

Scale range	1:25 000	1:50 000	1:100 000	1:200 000
Africa	2.9%	41.4%	21.7%	89.1%
Asia	15.2%	84%	56.4%	100%
Australia and Oceania	18.3%	24.3%	54.4%	100%
Europe	86.9%	96.2%	87.5%	90.9%
Former USSR	100%	100%	100%	100%
North America	54.1%	77.7%	37.3%	99.2%
South America	7%	33%	57.9%	84.4%
World	33.5%	65.6%	55.7%	95.1%

Table 1.2 Update rates of world mapping

Scale range	1:25 000	1:50 000	1:100 000	1:200 000
Africa	2.9%	41.4%	21.7%	89.1%
Asia	15.2%	84%	56.4%	100%
Australia and Oceania	18.3%	24.3%	54.4%	100%
Europe	86.9%	96.2%	87.5%	90.9%
Former USSR	100%	100%	100%	100%
North America	54.1%	77.7%	37.3%	99.2%
South America	7%	33%	57.9%	84.4%
World	33.5%	65.6%	55.7%	95.1%

photogrammetry applied in the twentieth century was able to provide mapping coverages in other continents, at least in priority areas. But technological, financial or organizational constraints have not been able to provide an updated basic mapping coverage needed for sustainable development. Therefore, assistance by new geoinformation technologies is required to achieve the necessary progress.

Integration of geoinformation technologies

Figure 1.2 outlines the differences between the classical spatial information system approach and the new geospatial information concept.

In the past, the disciplines of geodesy, photogrammetry and cartography worked in an independent fashion to provide printable maps. In the new concept GPS, remote sensing, digital photogrammetry and GIS are able not only to produce printable maps, but to display raster and vector images on a computer screen and to analyse them in an interdisciplinary manner for the purposes of society (see Figure 1.2).

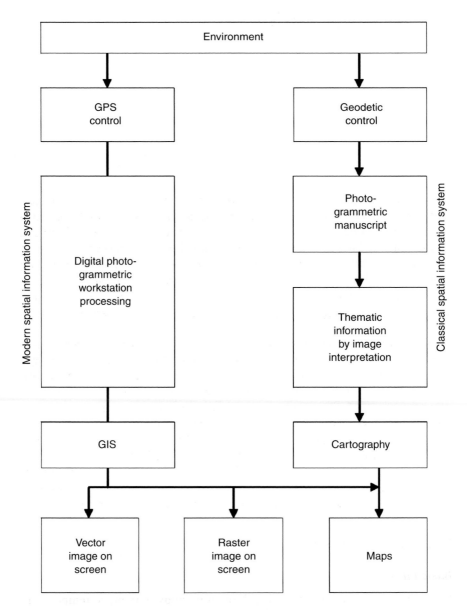

Figure 1.2 Classical and modern geospatial information system.

2 Remote sensing

Remote sensing is a method of obtaining information from distant objects without direct contact. This is possible due to the existence or the generation of force fields between the sensing device and the sensed object. Usable force fields are mechanical waves in solid matter (seismology) or in liquids (sound waves). But the principal force field used in remote sensing is that of electromagnetic energy, as characterized by the Maxwell equations. The emission of electromagnetic waves is suitable for directional separation. Thus, images of the radiation incident on a sensor may be generated and analysed.

The remote sensing principle using waves of the electromagnetic spectrum is illustrated in Figure 2.1. The energy radiates from an energy source. A passive (naturally available) energy source is the sun. An active energy source may be a lamp, a laser or a microwave transmitter with its antenna. The radiation propagates through a vacuum with the speed of light, c, at about 300 000 km/second. It reaches an object, where it interacts with the matter of this object. Part of the energy is reflected toward the sensor. At the sensor carried on a platform, the intensity of the incoming radiation is quantized and stored. The stored energy values are transformed into images, which may be subjected to image processing techniques before they are analysed to obtain object information.

Electromagnetic radiation

Basic laws

Electromagnetic energy is radiated by any body having a temperature higher than $-273°C$ (or $0°K$), the absolute zero temperature. Such a body radiates energy in all frequencies. The relation between frequency, v, and wavelength, λ, is expressible as

$$\lambda = \frac{c}{v},$$

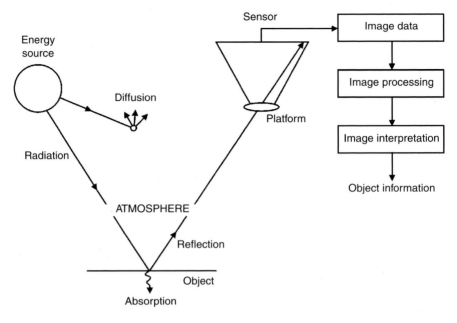

Figure 2.1 Principles of remote sensing.

with λ expressed in m and frequency in cycles/sec = Hertz.

The amount of radiation for a particular wavelength over an interval, $\Delta\lambda$, is a function of the absolute temperature of the body in °K and is expressed by Planck's distribution law:

$$L_\lambda(T) = \frac{2h \cdot c^2}{\lambda^5} \cdot \frac{1}{e^{h \cdot c/\lambda \cdot k \cdot T} - 1}$$

in which

$L_\lambda \cdot (T)$ is the spectral radiance in $W/m^2 \cdot sr \cdot \mu m$
$k = 1.38047 \cdot 10^{-23}$ $[w \cdot s \cdot K^{-1}]$, the Boltzmann constant
$h = 6.6252 \cdot 10^{-34}$ $[J \cdot s]$, the Planck's constant
$c = 2.997925 \cdot cm^8/s$

This is based upon the observation that energy is emitted in energy quantums, Q_P $[J = W \cdot s]$. Wien's law permits us to determine the radiation maximum at a particular wavelength, λ_{max}, depending on the body temperature by differentiating $L_\lambda(T)$ with respect to λ:

$$\lambda_{max} = \frac{0.002898}{T} \cdot \frac{[mL]}{[K]}$$

Since the surface temperature of the sun is about 6000°K, this means that the maximum radiance from solar energy will be generated at a wavelength of 480 nm, which corresponds to green light. The earth, with its temperature between 273°K and 300°K (0°C and 27°C) has its radiance maximum in the thermal range of the spectrum (8 μm to 14 μm).

Radiometric quantities

The dimensions of radiometric quantities used in absolute remote sensing are shown in Table 2.1.

The electromagnetic spectrum

The characteristics of electromagnetic energy in the electromagnetic spectrum are shown in Table 2.2.

Table 2.1 Radiometric quantities

	Name	*Symbol*	*Relation*	*Dimension*	
Radiator	Radiant energy	Q	–	Joule	$J = W \cdot s$
	Radiant flux	Φ	$\Phi = \dfrac{Q}{t}$	Watt	$W = J/s$
	Radiant intensity	I	$I = \dfrac{Q}{t \cdot \omega_1}$	–	W/sr
	Radiance	L	$L = \dfrac{Q}{t \cdot A_1 \omega_1}$	–	$W/sr \cdot m^2$
Emitted object	Radiant emittance	E	$E = \dfrac{\Phi}{A_1}$	–	W/m^2
Received signal	Irradiance	E′	$E' = \dfrac{\Phi}{A_2}$	–	W/m^2
	Exposure	H	$H = E' \cdot t$	–	$W \cdot s/m^2$

Notes
ω_1 = space angle of radiation in sterad (sr).
A_1 = radiating surface in m².
A_2 = irradiated surface in m².

Table 2.2 The electromagnetic spectrum

Radiation type	Wavelength	Frequency	Transmission	Use	Detector
Cosmic rays	10^{-13} to 10^{-16} m	4.7×10^{21} Hz to 3×10^{24} Hz	outer space	–	ionization detector
γ-rays	10^{-4} nm to 0.4 mm ($nm = 10^{-9}$ m)	8×10^{16} to 4.7 $\times 10^{21}$ Hz	limited through atmosphere	radioactivity	ionization detector
X-rays	0.4–10 nm	3×10^{16} to 8×10^{16} Hz	only at close range outer space	close range	phosphorus
Ultra-violet light	10–380 nm	7.9×10^{14} to 3×10^{16} Hz	weak through atmosphere	–	phosphorus
Visible light	380–780 nm	3.8×10^{14} to 7.9×10^{14} Hz	well through atmosphere	passive remote sensing, vision	photography, photodiode
Near infrared	780 nm–1 μm ($m = 10^{-6}$ km)	3.0×10^{24} to 3.8×10^{14} Hz	well through atmosphere	passive remote sensing	photography, photodiode
Medium infrared	1–8 μm	3.7×10^{13} to 3.0×10^{14} Hz	in windows through atmosphere	passive remote sensing	quantum detector
Thermal infrared	8 μm–1 mm ($mm = 10^{-3}$ m)	3×10^{11} to 3.7×10^{13} Hz	in windows through atmosphere, day and night	passive remote sensing	quantum detector
Thermal infrared	8 μm–14 μm (to 1 mm) ($mm = 10^{-3}$ m)	3×10^{11} to 3.7×10^{13} Hz	in windows through atmosphere, day and night	passive remote sensing	quantum detector
Microwaves	1 mm–1 m	300 MHz to 300 GHz (MHz = 10^6 Hz, GHz = 10^9 Hz)	day and night through clouds	active remote sensing	antenna
FM radio	1–10 m	30–300 MHz	direct visibility	TV, broadcast	antenna
Short-wave radio	10–100 m	3–30 MHz	worldwide	broadcast	antenna
Medium-wave broadcast	182 m–1 km	300–1650 KHz (KHz = 10^3 Hz)	regional	broadcast	antenna
Long-wave broadcast	1–10 km	30–300 KHz	worldwide	broadcast	antenna
Sound transmission	15–6000 km	50 Hz–20 KHz	cable	telephone	cable
AC	6000 km	50 Hz	cable	energy transmission	cable

Energy–matter interaction

Atmospheric transmission

Electromagnetic transmission through the atmosphere slows down the wave propagation depending on the transmission coefficient, n:

$$\lambda = \frac{c}{v \cdot n}$$

Furthermore, part of the energy is absorbed, when the energy quantums hit the molecules and atoms of the atmospheric gases.

Another part is directionally reflected into diffused energy causing scattering, according to Rayleigh.

$$I_{\text{transmitted}} = I_{\text{original}} \cdot \exp(-k \cdot r)$$

in which

k is the extinction coefficient and
r the distance passed through.
k is proportional to λ^{-4}.

In the infrared ranges of the spectrum, absorption takes place by gases contained in the atmosphere, so that transmission is possible only in atmospheric windows, as shown in Figure 2.2.

Energy interaction at the object

Interaction of incident electromagnetic energy with matter depends on the molecular and atomic structure of the object. Energy may be directionally

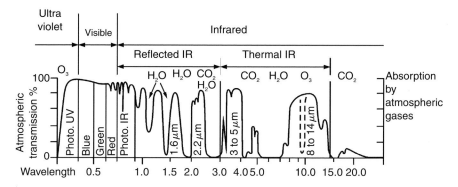

Figure 2.2 Atmospheric windows.

reflected, scattered, transmitted or absorbed. The process is caused by the interaction of a photon with an electron located in a shell of an atom, which results in the excitation of the electron from the shell. The ratio between the reflected (in all directions), transmitted and absorbed fluxes or radiances and the incoming radiation is described as:

reflection coefficient, ρ_λ
transmission coefficient, τ_λ
and absorption coefficient, α_λ.

Their sum is equal to 1:

$$\rho_\lambda = \frac{\phi_{\lambda\,reflected}}{\phi_{\lambda\,incoming}}$$

$$\tau_\lambda = \frac{\phi_{\lambda\,transmitted}}{\phi_{\lambda\,incoming}}$$

$$\rho_\lambda = \frac{\phi_{\lambda\,absorbed}}{\phi_{\lambda\,incoming}}$$

$$\rho_\lambda + \tau_\lambda + \alpha_\lambda = 1$$

Other than through atmospheric particles, transmission is possible, for example, for water. Non-transparent solid bodies have a transmission coefficient of 0.

The incoming energy cannot get lost. Absorption is a process in which higher frequency energy (e.g. light) is converted to lower frequency energy (e.g. heat).

The reflection coefficient of an object is of crucial importance for remote sensing. It varies for different spectral ranges for a particular object. It is characterized by an angle dependent function, the so-called radiometric function:

$$\rho_\lambda(\epsilon_1, \epsilon_2, \epsilon_3) = \rho_{\lambda_0} \cdot f(\epsilon_1, \epsilon_2, \epsilon_3) = \rho_\lambda \text{ (solar position)}$$

in which ρ_{λ_0} is the normal reflection coefficient, valid for an object illuminated and reflected in the same direction. ϵ_1 is the spatial angle between the direction of illumination and the surface normal. ϵ_2 is the spatial angle between the direction of the sensor and the surface normal. ϵ_3 is the spatial angle between the direction of illumination and the direction to the sensor.

The normal reflection coefficients vary for different object types, e.g. for green light:

coniferous forest	1 per cent
water	3 per cent
meadow	7 per cent
road	8 per cent
deciduous forest	18 per cent
sand	25 per cent
limestone	60 per cent
new snow	78 per cent.

The spectral reflectance varies with each object type, as shown in Figure 2.3 for soil and vegetation.

Interaction of energy in plants

While the reflection on smooth object surfaces is simple, the interaction of energy in plants is more complicated. A plant leaf, according to Figure 2.4, consists of at least three layers:

- transparent epidermis.
- a palisade type mesophyll, which reflects green light and absorbs red, due to its chlorophyll content.
- a spongy type mesophyll which reflects near infrared.

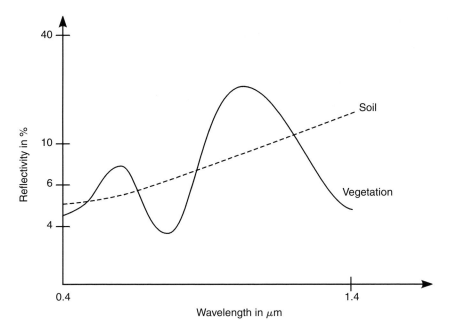

Figure 2.3 Spectral reflectance.

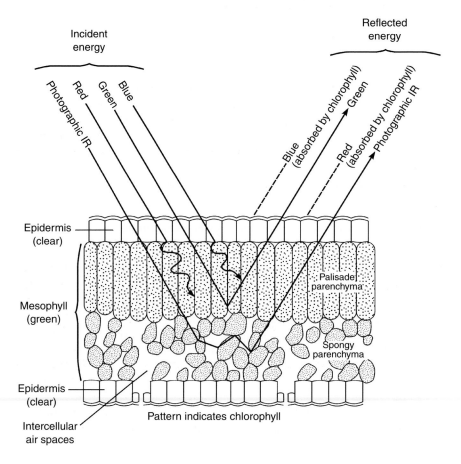

Figure 2.4 Cross-section of a leaf.

Energy arriving at the sensor

The radiant density flux arriving at a sensor is composed of the following terms: the radiant flux of solar radiation, E_λ, is diminished by the solar altitude $(90° - \epsilon_1)$ and the transmission coefficient of the atmosphere $\tau_{\lambda\epsilon_1}$ for the incident ray to result in the radiant flux of the object E_{ϵ_1}:

$$E_{\epsilon_1} = \cos \epsilon_1 \int_{\lambda_1}^{\lambda_2} E_\lambda \cdot \tau_{\lambda\epsilon_1} \cdot d_\lambda$$

The reflection at the object is dependent on the radiometric function. In case the object is a Lambert reflector, reflecting all incident energy into the half sphere with equal intensity, $f(\epsilon_1, \epsilon_2, \epsilon_3)$ becomes equal to 1.

The radiant flux arriving at the sensor then becomes:

$$E_{\epsilon_2} = \frac{\cos \epsilon_1 1}{\pi} \int_{\lambda_1}^{\lambda_2} E_\lambda \cdot \tau_{\lambda\epsilon_1} \cdot \tau_{\lambda\epsilon_2} \cdot \rho_\lambda \cdot d\lambda$$

with $\tau_{\lambda\epsilon_2}$ as transmission coefficient of the atmosphere between object and sensor and the reflection coefficient, ρ_λ, equal in all directions of reflection.

The actual radiant flux arriving at the sensor, as far as its intensity is concerned, is still augmented, however, by the diffused light of the atmosphere $\rho_{\epsilon_1\lambda}$, which amounts to about 3 per cent. It is strongly wavelength dependent.

Thus, the incident radial flux, E, at the sensor becomes:

$$E = \frac{1}{\pi} \int_{\lambda_1}^{\lambda_2} E_\lambda (\tau_{\lambda\epsilon_1} \cdot \tau_{\lambda\epsilon_2} \cdot \rho_\lambda \cdot \cos \epsilon_1 + \rho_{\epsilon_1\lambda}) d\lambda$$

E_λ can be derived from solar parameters. The transmission coefficients of the atmosphere can be determined from radiometric measurements of the solar illumination with radiometers directed at the sun and the total illumination of the half sphere, measurable by an Ulbricht sphere in front of the radiometer. With ϵ_1 known and $\rho_{\epsilon_1\lambda}$ estimated for the diffused light, remote sensing can be treated as a procedure of absolute radiometry.

To use this absolute methodology is, however, impractical because of the need to measure atmospheric transmission parameters and because of assumptions for the radiometric function and the scattered energy.

Thus, remote sensing generally restricts itself to a relative comparison of directionally separable radiant fluxes or radiances of adjacent objects A and B:

$$\frac{E_A}{E_B} = \frac{L_A}{L_B} = \frac{\rho_A}{\rho_B}$$

as E_λ, $\tau_{\lambda\epsilon_1}$, $\cos \epsilon_1$ and $\rho_{\epsilon_1\lambda}$ are nearly equal in this case.

Sensor components

The classical remote sensing device is photography.

Optical imaging

The first historical device to create an image of a scene was the pinhole camera. It contains a small circular hole, which can be considered as the projection centre. Light rays passing through this hole may be imaged in an image plane at an arbitrary distance.

The disadvantage of the pinhole camera is that the incident radiation in the image plane is too weak for practical purposes. Optical lenses permit the collection of more radiation. To be imaged the lenses must, however, focus the image according to the condition:

$$\frac{1}{a} + \frac{1}{b} = \frac{1}{f}$$

in which a is the distance between projection centre and object point, and b is the distance between projection centre and image point, and f is the focal length of the spherically shaped lens.

Simple lenses suffer a number of sharpness restrictions, such as:

- chromatic aberration, which focuses light of different wavelengths in separate image planes.
- spherical aberrations, which collect the inner rays of the lens in different planes with respect to the outer rays of the lens.
- astigmatism, which causes the rays passing vertically through the lens to be collected in a different plane than for the horizontal rays.
- curvature of the field causes the outer rays and the inner rays passing through the lens not in a plane but on a spherical surface (see Figure 2.5).

Lens manufacturers have therefore attempted to combine lenses of different refractive indices, correcting the sharpness deficiencies of simple lenses.

The amount of energy arriving in the image plane through a lens system with the diaphragm, d, is a function of the angle α between its optical axis and the imaged point. This function is influenced by the longer path of the ray, the transmission coefficient of the lens system and the diameter. For a single lens, the irradiance diminishes with $\cos^4\alpha$. Objectives are able to reduce the light fall off to a function $\cos^{2.5}\alpha$.

The exposure, H, at a point on the optical axis of the image plane thus becomes:

$$H = E \cdot \left(\frac{d}{f}\right)^2 \cdot \frac{\pi}{4} \cdot \tau_0 \cdot t$$

with E being the irradiant radial flux at the sensor, d the diaphragm of the optics, f the focal length, t the exposure time and τ_0 the transmission coefficient of the objective in the direction of the object, which is α dependent.

The photographic process

The optically produced image in the image plane can be recovered by the

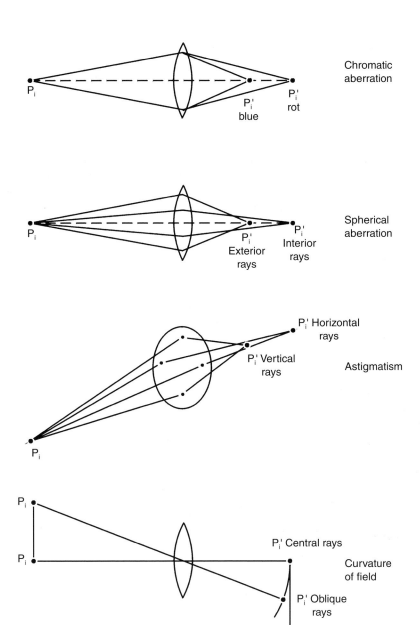

Chromatic aberration

P_i P_i' blue P_i' rot

Spherical aberration

P_i P_i' Exterior rays P_i' Interior rays

Astigmatism

P_i' Horizontal rays P_i' Vertical rays P_i

Curvature of field

P_i P_i P_i' Central rays P_i' Oblique rays

Figure 2.5 Lens errors.

photographic process on a film. A layer of film consists of silver halides embedded in gelatine forming the emulsion. When light falls on the emulsion, the silver halide particles are in part chemically converted into metallic silver. The amount of this conversion is proportional to the irradiance.

During the wet development process in the laboratory by methylhydrochinon, the conversion process is continued. A second wet process, the fixation, stops the conversion and dissolves the remaining silver halides. After a washing process, the metallic silver becomes visible. Thereafter the film is subjected to a drying process in which a copiable film negative is produced. It may be copied by illumination onto a copying film, which is subjected to the same process of development, fixation, washing and drying to produce a diapositive.

The grey level of the produced metallic silver on negative or diapositive corresponds to a logarithmic function of the irradiance. This is shown in Figure 2.6 as the so-called 'D-logE-curve', which should better be called 'D-logH-curve' linking photographic density to irradiance.

The D-logH-curve is characteristic for a particular film type. Exposure of the entire irradiance range of the scene should be chosen within the linear range of the D-logH-curve. In this range, the slope of the curve is equal to the gradation γ:

$$tg\alpha = \frac{\Delta D}{\Delta \log H} = \gamma$$

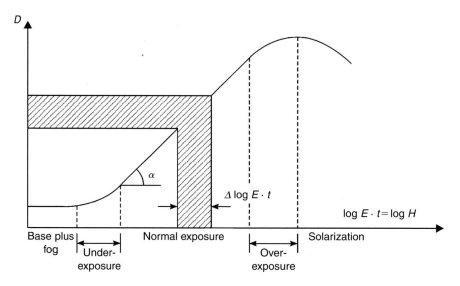

Figure 2.6 D-logH-curve.

The gradation of the film can be influenced by the development process.

$\gamma < 1$ is called a soft development.
$\gamma = 1$ is a normal development.
$\gamma >$ is a hard development.

γ can be changed by the choice of the photographic material, by the developer type, by the temperature of the developer and by the duration of the development process.

If the development process is fixed as a standard for the type of developer, temperature and the duration of development, then the sensitivity of the film material can be defined (see Figure 2.7).

A point *A* on the density curve is reached when the density difference ΔD above fog is 0.1. For this point, the sensitivity in ASA or ISO is

$$\frac{0.8}{H_A}$$

and the sensitivity in DIN is

$$10 \cdot \log \frac{1}{H_A}$$

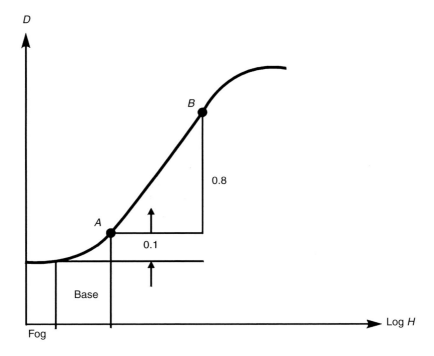

Figure 2.7 Photographic sensitivity.

The development process is fixed as a standard so that an irradiance difference

$$\Delta \log H_{AB} = 1.30$$

corresponds to a density difference

$$\Delta D_{AB} = 0.8$$

Silver halides are most sensitive to blue light, which is heavily influenced by scattering in the atmosphere. If this scattering effect is to be minimized, imaging through a yellow filter is mandatory.

The addition of optical dyes permits us to extend the sensitivity of a film into the near infrared region, not visible to the eye, in which vegetation reflects heavily. Thus special film types have become available for the visible spectrum or for the spectral range from green to infrared, filtering out the blue light (see Figure 2.8).

The grey value of the negative or diapositive is determined by the transparency of the developed emulsion. Transparency is the ratio between the light flux passing the emulsion ϕ and the incoming light flux ϕ_0:

$$\tau = \frac{\phi}{\phi_0}.$$

$1/\tau$ is called the opacity, density D is defined as:

$$D = \log \frac{1}{\tau}.$$

Colour and colour photography

As human vision has the ability to distinguish different frequency components of the radiances of objects in the form of colour, a definition of the colour system has been necessary. The Commission Internationale de l'Eclairage (CIE) has defined the three principal colours for the following wavelengths:

blue $= 435.8$ nm.
green $= 546.1$ nm.
red $= 700.0$ nm.

Each perceived colour corresponds to an addition of the three principal colour components. All colours can be represented in the CIE chromaticity diagram shown in Figure 2.9.

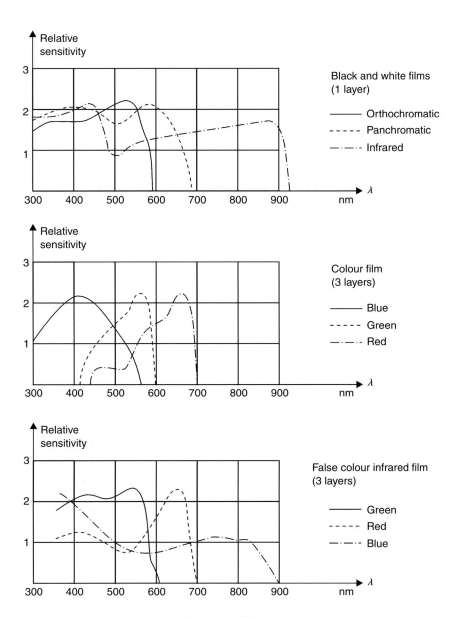

Figure 2.8 Spectral sensitivity of different films.

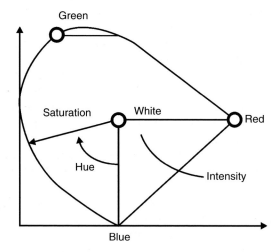

Figure 2.9 Chromaticity diagram.

If the colours red, green and blue are equally represented, their mixing appears to the eye as white at a particular intensity (brightness). On the outside curve of the diagram are pure colours corresponding to a particular frequency between 380 nm and 770 nm. These represent the hue (dominant wavelength) of the light received. Inside the diagram a particular light, coming from a mix of frequencies, is represented by the saturation (purity) of a colour.

The colours may be generated artificially by the projection of images of different grey levels in the three primary colours (colour additive process). This is the case for a computer or television screen, where three images filtered for the three primary colours are added.

In most photographic work, a colour subtractive process is used by absorption filters. This is based on the subtraction of a colour from white:

white − red = cyan.
white − green = magenta.
white − blue = yellow.

In colour or false colour films, three separate film layers are used, which have been sensitized and filtered for the three principal spectral ranges. A colour film has a blue sensitive layer followed by a yellow filter, then a green sensitive layer and finally a red sensitive layer attached to a film base.

Let us now look at the exposure and development process. For colour reversal films producing a diapositive, the exposure with blue, green and red light will initiate the creation of metallic silver in the respective layers. The film is subjected to a black and white development. This develops the metallic silver in the respective layers. The result is a black and white negative, which is still light sensitive in the unexposed layers. A short subsequent illumination of the film will therefore create silver in the previously unexposed layers. This silver is colour-developed with colour dyes in complementary colours (cyan for red, magenta for green and yellow for blue). Bleaching of the film will convert all non-colour coupled metallic silver into soluble silver salts. Thus a film will be the result, which in transparent viewing will yield the original object colours.

Colour negative films possess the same three layers, which expose metallic silver in the respective layers. However, the generated silver is directly developed with the colour dyes in complementary colours, creating the oxidization product, which cannot be removed by the bleaching process. Thus a colour negative will result, which when copied by the same process can yield a colour image in the original colours.

False colour photography utilizes three layers sensitive to green, red and near infrared. The development process is similar to that of colour reversal film. The green sensitive layer appears blue in the resulting diapositive, the red sensitive layer in blue and the infrared sensitive layer in red. The false colour film is particularly useful in interpreting the health of vegetation.

Digital imaging

The irradiated element of a digital imaging system is a photo-diode. This charge-coupled device (CCD) consists of a metallic electrode on a silicon semiconductor separated by an oxide insulator. An incoming photon is attracted by the positive voltage of the electrode. A charge is thereby created at the semiconductor acting as a capacitor. The charge is proportional to the number of photons arriving at the electrode (see Figure 2.10).

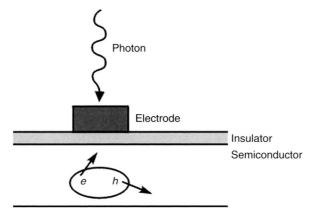

Figure 2.10 CCD detector.

CCDs are now available at a pixel size up to $5\,\mu m$. They may be grouped into linear arrays.

To read out the charges of the diodes of a linear array, a voltage is applied, so that the charge of one diode is shifted to the next one along. This process is repeated until the charges of the whole line are shifted to a storage array as an analogue video signal (see Figure 2.11).

This video signal can be captured using a tuning signal up to thirty times per second by an analogue to digital converter, called a 'frame grabber'. The frame grabber captures the exposure as a digital grey level value for later digital processing.

CCD area sensors are composed of several linear arrays arranged in a matrix fashion. The readout into a storage zone occurs line by line in parallel. While the signal is digitally converted in the frame grabber, a new image can be exposed.

The resolution of CCD sensors is limited by the sensor area and by the distance between the sensor elements. These determine a maximal sampling frequency. The geometrical properties of the sensor are governed by the precision with which the sensor elements may be positioned in the image plane.

In contrast to photography, which has a logarithmic D-logH-curve, digital sensors have a linear D-H response. As a rule, the digital quantization of the grey level signals is in 2^8 or 256 bit. Newer sensors permit quantization to 2^{11} or 2048 grey levels.

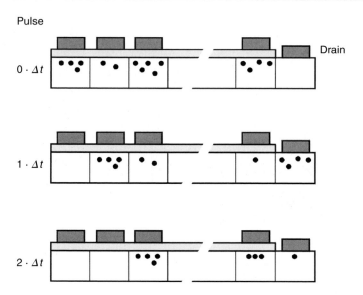

Figure 2.11 Charge transfer in a CCD array.

Imaging sensors

Aerial survey cameras

Aerial survey cameras must meet stringent requirements to reach high resolution and low geometric distortion for the image. This is achieved through the design of a suitable objective, as shown in Figures 2.12 and 2.13.

The objective is placed into a vibration dampened mount in a vertical hole in the floor of an aircraft. The camera has a rigid frame in the image plane carrying the fiducial marks, which permit the location of the principal point of the camera. The knowledge of the principal point in the image plane is important, as it defines the interior orientation for the application of the perspective laws according to which an image can be geometrically restituted. The relation of an objective with respect to the image plane is shown in Figure 2.14.

The aerial film of up to 120 m length and 24 cm width is contained in a cassette mounted on top of the image plane. It contains a pressure plate, which at the time of the exposure is pressed against the image frame. At that time, the film guided through the image plane is pressed onto the pressure plate by a vacuum system to assure the flatness of the exposed film. For that purpose, the pressure plate is equipped with suction holes. After the exposure, the film pressure is released to allow forward motion of the film for the next exposure by about 25 cm, accounting for the standardized image format of 23 × 23 cm.

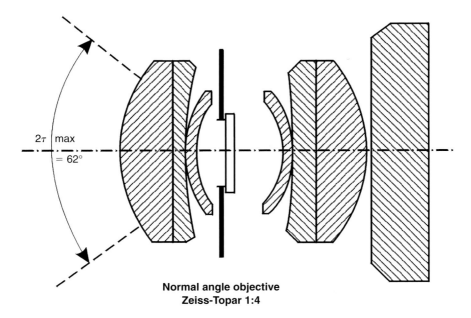

Normal angle objective
Zeiss-Topar 1:4

Figure 2.12 Normal angle objective.

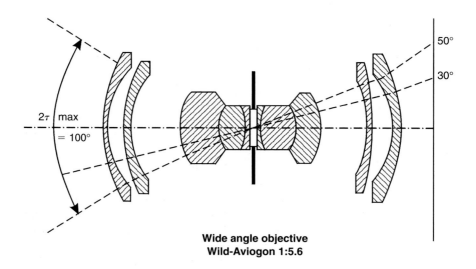

**Wide angle objective
Wild-Aviogon 1:5.6**

Figure 2.13 Wide angle objective.

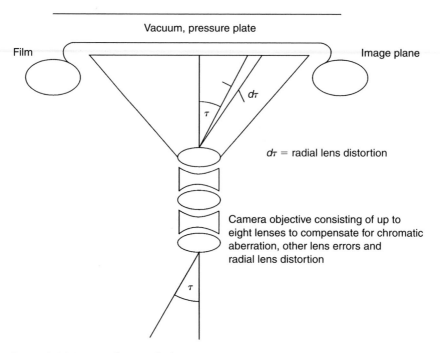

Figure 2.14 Image plane and objective.

Photography is controlled by a rotating shutter, permitting a simultaneous exposure of all parts of the image every 2 seconds for a duration between 1/100 and 1/1000 of a second. The more recent camera types (Leica RC 30 and Zeiss RMK TOP) permit a forward motion of the film during the exposure to enable the exposure of the same terrain for longer intervals, while the aeroplane moves forward, permitting the use of high-resolution film, which, due to its smaller silver halide grains, needs a longer exposure time for an exposure falling into the linear part of the D-logH-curve. This image motion compensation generally improves the achievable resolution (see Figure 2.15).

Figures 2.16 and 2.17 show the aerial camera types produced by the manufacturers LH-Systems and Z/I Imaging.

A systematic survey of the terrain is only possible if the camera frame is oriented in parallel in the flight direction. To eliminate the crab angle between camera orientation and the flight axis, the camera is rotated within the mount by a mechanism (see Figure 2.18).

Older cameras control this movement by the flight photographer using an overlap regulator consisting of a ground glass image plane onto which the ground is continuously imaged while the aircraft moves forward. The overlap regulator permits the control of exposure interval of more than 2 seconds, so that a particular overlap of ordinarily 60 per cent in the flight

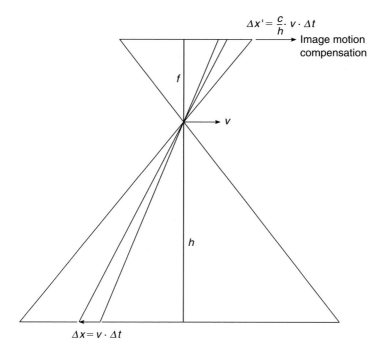

Figure 2.15 Image motion compensation.

Figure 2.16 The LH Systems RC30 camera.

Source: Image courtesy of LH Systems (Leica Geosystems), San Diego, CA. © Leica Geosystems, 2000. LH Systems is a wholly owned subsidiary of Leica Geosystems and produces airborne data acquisition systems within the Leica Geosystems' GIS & Mapping Division.

Figure 2.17 The Z/I RMK TOP camera.

Source: Image courtesy of Z/I Imaging Corp., Oberkochen.

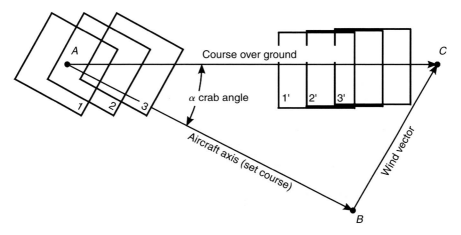

Figure 2.18 Crab angle.

direction is achieved. Newer cameras control orientation of the camera mount by the crab angle and the set overlap percentage by an automatic sensor, which checks the aircraft motion over the ground *h/v*. The time interval between exposures, *t*, for a given overlap, *q*, is calculated by:

$$t = \frac{s}{v} = \frac{h}{v} \cdot \frac{a'}{f}\left(1 - \frac{q}{100}\right)$$

in which *s* is the ground distance between exposures, *v* the velocity of the aircraft, *a'* the image size (23 cm) and *f* the chosen focal length (equal to the principal distance for imaging at infinity).

The major aerial survey camera types in use are contained in Table 2.3, along with their characteristics.

Table 2.3 Analogue camera systems

Manufacturer	Name of camera	Type of objective	Image angle $2\tau_{max}$	Principal distance f	Ratio d/f	Maximal image distortion
LH Systems	RC 30	UAG-S	90°	153 mm	1:4	4 μm
LH Systems	RC 30	NATS-S	55°	300 mm	1:5.6	4 μm
Z/I Imaging	RMK-TOP 15	Pleogon A3	93°	153 mm	1:4	3 μm
Z/I Imaging	RMK-TOP 30	TOPAR A3	54°	305 mm	1:5.6	3 μm

Image quality of aerial survey cameras

Image quality in photographic images is determined by the contrast between two image objects with the irradiances E_0 and E_N or its reflection coefficients ρ_o and ρ_N. The object contrast is therefore:

$$K = \frac{E_N}{E_0} = \frac{\rho_N}{\rho_o}$$

Modulation is defined by:

$$M = \frac{\rho_N - \rho_o}{\rho_N + \rho_o} = \frac{K - 1}{K + 1}$$

This object contrast or the corresponding object modulation is visible in the image as image contrast K' or as image modulation M' depending on the transparencies τ_o and τ_N of the imaged object contrast

$$K' = \frac{\tau_N}{\tau_o}$$

and

$$M' = \frac{\tau_N - \tau_o}{\tau_N + \tau_o} = \frac{K' - 1}{K' + 1} = \frac{10^{\Delta D} - 1}{10^{\Delta D} + 1}$$

with

$$\Delta D = \log \frac{\tau_N}{\tau_o}$$

Contrast density difference and modulation can therefore be converted into corresponding values shown in Table 2.4.

The contrast transfer, C', and the modulation transfer, C, through the imaging system is therefore:

$$C' = \frac{K'}{K}, \text{ and } C = \frac{M'}{M}$$

Table 2.4 Contrast, density difference and modulation

Contrast	1000:1	2:1	1.6:1
Density difference	3.0	0.3	0.2
Modulation	0.999	0.33	0.23

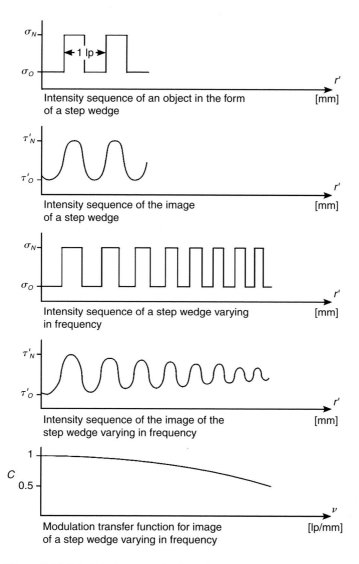

Intensity sequence of an object in the form of a step wedge [mm]

Intensity sequence of the image of a step wedge [mm]

Intensity sequence of a step wedge varying in frequency [mm]

Intensity sequence of the image of the step wedge varying in frequency [mm]

Modulation transfer function for image of a step wedge varying in frequency [lp/mm]

Figure 2.19 Modulation transfer function.

Both C' and C are functions of the spatial distance, expressed in terms of a spatial frequency ν. This is shown in Figure 2.19 for the frequency dependent modulation transfer function.

Image quality in a combined optical–photographic system is deteriorated by the imaging process if the modulation transfer function, C, is <1. Only for an extremely high contrast (1000:1) does C become nearly 1.

Image quality in a photographic system, other than by contrast, is influenced by the following components:

- the optical system.
- the film.
- the image motion during exposure.

Each component is characterized by its own modulation transfer function, C_{optics}, C_{film}, $C_{image\ motion}$. The total effect is C_{total}:

$$C_{total} = C_{optics} \cdot C_{film} \cdot C_{image\ motion}$$

The typical components for an aerial camera system are shown in Figure 2.20.

The resolution of an optical system is first limited by diffraction. The resolution A in lp/mm can be expressed by:

$$A = \frac{1000 \cdot d}{24 \cdot \lambda \cdot f}$$

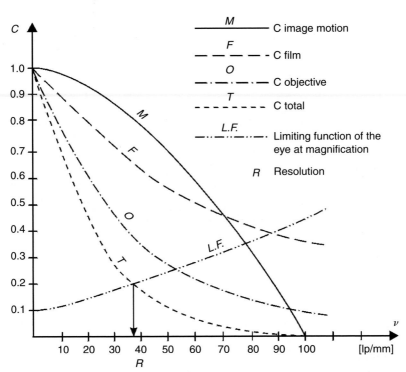

Figure 2.20 Modulation transfer function components for an aerial photographic system.

in which d is the diameter of the diaphragm, λ the wavelength and f the focal length. For a d/f value of 4, this amounts to 120 lp/mm. Other limiting effects stem from the composition of the lens system causing lower actual resolution.

The photographic film resolution depends on its grain size. High sensitive film contains coarse silver halide particles with a resulting low resolution. Low sensitive film has small grain size of the particles resulting in high resolution. The resolution of typical aerial films is shown in Table 2.5.

Image motion $\Delta a'$ in micrometres during exposure depends on the exposure time, Δt, on the platform movement over the ground v in kilometres per hour, the focal length f and the flight altitude h:

$$\Delta a' = \frac{f}{h} \cdot \frac{\Delta t}{3600} \cdot v$$

Under the assumption that the modulation assumes a value of 0 at a value of

$$a' = \frac{1000}{\Delta a'}$$

in lp/mm, the modulation transfer function due to image motion becomes:

$$C_{\text{image motion}} = \cos\left(\frac{v}{2a'}\right)$$

The human eye determines a limiting function for which contrasts can still be recognized at low modulation. Intersecting this limiting function with C_{total} yields the interpretable resolution in lp/mm.

The atmosphere generally diminishes the object contrast for all objects sensed by aerial imaging. If two objects consisting of sand with a reflection coefficient $\rho_{\text{sand}} = 30$ and coniferous forest with a reflection coefficient $\rho_{\text{forest}} = 1$ are observed, the object contrast, K, may be expressed as:

$$\frac{\rho_{\text{sand}}}{\rho_{\text{forest}}} = \frac{30}{1} = 30 = K$$

The image contrast is then affected by the diffused atmospheric light of 3 per cent.

Thus:

$$K' = \frac{30 + 3}{1 + 3} = \frac{33}{4} = 8 = K'$$

This illustrates that resolutions of generally low contrast aerial

Table 2.5 Aerial survey films

Manufacturer	Name	Sensitivity	Resolution of contrast 1000:1	Resolution at contrast 1.6:1	Type
Agfa	Aviphot Pan 200 S PE1	125–200 C ISO (23 DIN)	130lp/mm	50lp/mm	panchromatic
Agfa	Aviphot pan 80	64–100 ISO (20 DIN)	287lp/mm	101lp/mm	panchromatic
Agfa	Aviphot color X100 PE	125–160 ISO (22 DIN)	140lp/mm	55lp/mm	colour negative
Agfa	Aviphot color N400 PE	400–640 ISO (28 DIN)	130lp/mm	35lp/mm	colour negative
Agfa	Aviphot chrome 200 PE1/PE3	200 C ISO (24 DIN)	110lp/mm	50lp/mm	colour reversal
Kodak	Aerographic 2402	160 ISO (23 DIN)	130lp/mm	55lp/mm	panchromatic
Kodak	Aerographic 2403	640 ISO (29 DIN)	100lp/mm	40lp/mm	panchromatic
Kodak	Double X Aerographic 2405	400 ISO (27 DIN)	125lp/mm	50lp/mm	panchromatic
Kodak	Panatomic X 3412	40 ISO (17 DIN)	400lp/mm	125lp/mm	panchromatic
Kodak	Aerocolor 2445	64 ISO (19 DIN)	80lp/mm	40lp/mm	colour negative
Kodak	Aerochrome 2448	32 ISO (16 DIN)	80lp/mm	40lp/mm	colour reversal
Kodak	Aerocolor SO 846	160 ISO (23 DIN)	100lp/mm	63lp/mm	colour negative

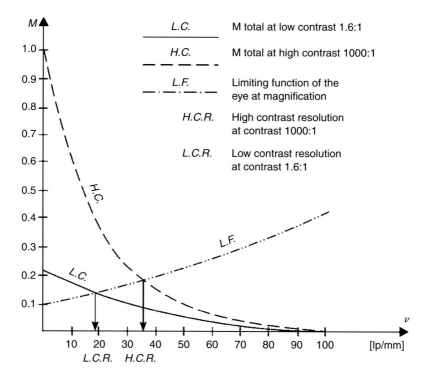

Figure 2.21 Resolution at low contrast.

photographs should not be compared at a contrast of 1000:1, but at a contrast of 1.6:1, for which the modulation of 1 diminishes to 0.23. Thus the modulation transfer function, as shown in Figure 2.21, is flattened, and its intersection with the limiting function of the eye is lowered in resolution.

Aerial survey camera systems with image motion compensation are, in practice, achieving a resolution of between 40 to 50 lp/mm for black and white images and of between 30 to 40 lp/mm for colour images.

Another property of aerial survey cameras is the ability to generate photographic images with minimum geometric distortion. A primary source of geometric distortion stems from the objective (see Figure 2.22).

Radial distortion, $\Delta \tau$, is the angular difference between the angle of exposure direction and object direction, τ_i, and their imaged difference τ_i':

$$\Delta \tau_i = \tau_i - \tau_i'$$

Distortion can also be expressed as a radial distance difference in the image plane

$$\Delta_r' = r_i - r_i'$$

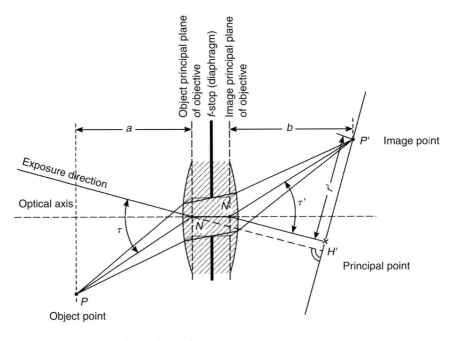

Figure 2.22 Imaging through an objective.

with r'_i measured from the principal point, determinable from the fiducial marks of the camera and $r_i = f \cdot tg\ \tau_I$ (see Figures 2.23 and 2.24). N marks the entrance of exit modes of rays along the optical axis through the objective.

Camera manufacturers have tried to minimize the distortion in a factory calibration procedure in which the angles τ'_i are measured through the objective by a goniometer. The fiducial marks of the camera, determining

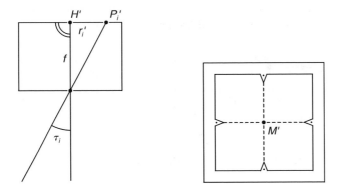

Figure 2.23 Principal point.

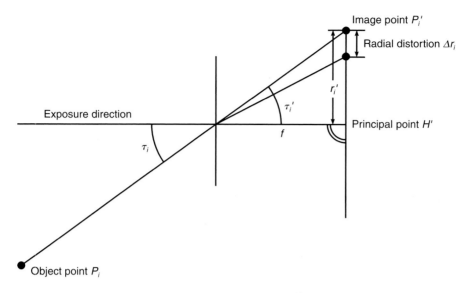

Figure 2.24 Definition of radial distortion.

the principal point, are fixed so that a minimum radial and tangential distortion results.

Camera manufacturers issue calibration certificates in which the attained radial distortions are listed. They are in the order of $2\,\mu m$ and usually never exceed $4\,\mu m$ in the image plane. Tangential lens distortions are generally only one-third of the radial distortions and thus negligible for factory calibrations (see Figure 2.25).

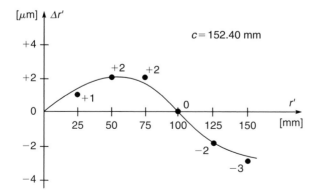

Figure 2.25 Radial distortion of a photogrammetric objective.

A second source of distortion is the film. If the pressure plate vacuum has properly worked, the image plane can be considered as flat. Otherwise distortions of

$$\Delta r'_i = \frac{r'}{f} \Delta f = \Delta f \cdot tg\tau'$$

will result. Δf is the deviation of the image point on the film from the ideal focal plane at the principal distance f.

More serious are the geometric distortions of the film caused by the wet developing and drying process. For this reason, each camera contains between four to eight fiducial marks. If their images are measured, a correction of the film deformations in the direction of the film and perpendicular to it can be made during the restitution. Fortunately, these types of film deformations remain constant for the entire film, so that they can be accounted for by parameters in the restitution process.

Opto-mechanical scanners

An opto-mechanical scanner contains a single sensor element, which permits the recording of the irradiance of a ground pixel. A rotating mirror scans the terrain, so that a whole line of ground pixels can be recorded in a time sequence. The next scan of a forward-moving platform records the adjacent line of ground pixels. Thus, the scanning mechanism during forward motion permits a recording of a whole image. The scanning principle is shown in Figure 2.26.

For an altitude, h, a field of view of the scanned pixel, ω, the total angular scan width, Ω, and the scan angle from the vertical, α, the ground pixel size, a, in the scan direction becomes:

$$a = \frac{\omega \cdot h}{\cos^2 \alpha}$$

The ground pixel size, b, in the line of flight, b, is likewise:

$$b = \frac{\omega \cdot h}{\cos^2 \alpha}$$

The swath of the scan, s, becomes:

$$s = 2h \cdot tg\frac{\Omega}{2}$$

The scan frequency, ν, is a function of the platform velocity, v_g:

$$\nu = \frac{v_g}{\omega \cdot h}$$

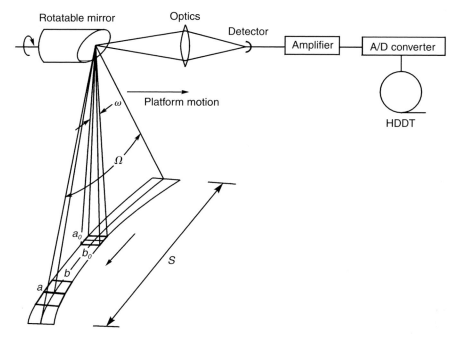

Figure 2.26 Operation of an opto-mechanical scanner.

Opto-mechanical scanners have been utilized for airborne and for satellite multispectral scanners, as shown in Figure 2.27.

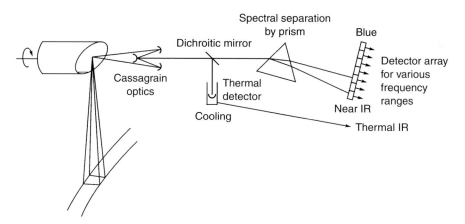

Figure 2.27 Operation of a multispectral scanner.

Instead of a single photodiode, a linear array was used as a sensor. The spectral separation of the incoming energy was achieved by the diffraction of a prism in the optical path, so that different bands of wavelengths could be recorded at the array at a particular time.

The scanner even permitted the recording of thermal energy with the use of a dichroitic separation of the ray. Thermal energy could then be collected on a thermally cooled (77°K or 5°K) far-infrared sensitive mercury-doped CD-telluride or germanium detector.

The characteristics of such detectors are shown in Table 2.6.

Laser scanners

It is possible to direct laser light impulses to the terrain along the principles of electro-mechanical scanning. Part of the reflected radiation returns to the laser scanner. This gives the possibility for measuring the time between the emission of the pulse and the first and the last return, as well as the energy received.

This permits the use of the device as a laser altimeter for any spot reached by the laser pulse. If the position of the sensor is measured by in-flight GPS, and its orientation by inertial navigation devices, then it becomes possible to determine the three-dimensional position of the reflection point.

On this principle laser scanners, such as the system operated by Toposys, have been built and operated: it is a pulsed fibre scanner, operated at an airborne altitude of less than 1600 m with a laser wavelength of 1.55 μm. It emits pulses every 5 nsec. The scan frequency is 650 Hz and the pulse repetition rate 83 000 Hz. This permits, within a field of view of 7° from the vertical, the measurement of laser reflection points of a density of three points per m^2 on the surface.

The recording of the first and last pulse received from the point permits the judging of the thickness of the vegetation cover of the terrain. The swath covered in a flight strip is then up to 390 m, and elevation measurements are possible within a relative accuracy of 2 cm and an absolute accuracy of 15 cm for a digital elevation model. Leica-Helava Systems produces

Table 2.6 Opto-mechanical detectors

Scanner	Platform	ω	Ω	Number of visual and near IR channels	Number of thermal channels
Daedalus DS 1200	aircraft	2.5 mrad	77°	2	1
Landsat MSS	satellite	0.087 mrad	11.6°	4	0
Landsat TM	satellite	0.024 mrad	11.6°	6	1

a laser scanner ALS 40 for the generation of digital elevation models. Figures 2.28 and 2.29 show the laser signals received in original and filtered forms.

Figure 2.28 Original laser scan.

Source: Image courtesy of Institute for Photogrammetry and GeoInformation, University of Hannover.

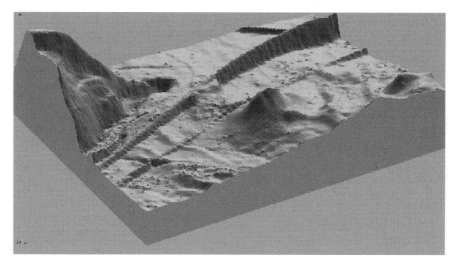

Figure 2.29 Filtered laser scan.

Source: Image courtesy of Institute for Photogrammetry and GeoInformation, University of Hannover.

It is possible to use the intensity of the received signal to construct an image with respect to the digital elevation model, even though this use of the laser scanner is still experimental (see www.toposys.com, www. geolas.com).

Opto-electronic scanners

Opto-electronic imaging is possible when an image created through an optical system is created on a linear or a matrix array of CCD sensors. Linear arrays of sufficient length are easier to assemble than matrix arrays. Therefore the electro-optical scanner, as shown in Figure 2.30, has frequently been applied in satellite sensors using the push-broom principle.

A push-broom scanner has the linear array oriented perpendicular to the platform motion. A detector element of the dimension a' perpendicular to the flight direction will cover a ground pixel dimension, a, according to:

$$a = \frac{h}{f} \cdot a'$$

with h being the platform altitude and f the focal length of the optics. In-

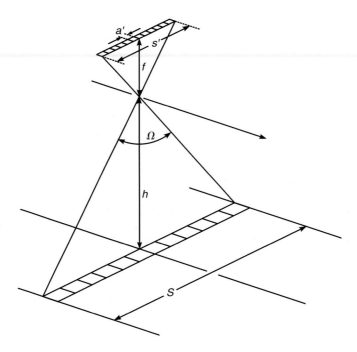

Figure 2.30 Opto-electronic scanner.

flight direction, the detector element of the dimension b' in this direction is likewise imaged at a ground pixel dimension, b:

$$b = \frac{f}{h} \cdot b'$$

To assemble an image, the exposure, Δt, must be chosen proportional to the velocity of the platform with respect to the ground, v_g:

$$b = v_g \cdot \Delta t$$

The swath of the push-broom sensor, s, becomes

$$s = \frac{h}{f} \cdot s' = 2h \cdot tg\frac{\Omega}{2}$$

with s' being the length of the linear array pointing symmetrically to the vertical. An example is the sensor for the French Spot satellite, characterized in Table 2.7.

The Spot satellite sensor not only uses a panchromatic array with 10 m ground pixels but, in parallel to that, three further arrays of half resolution, yielding 20 m ground pixels for the filtered spectral bands of green, red and near infrared.

One Spot satellite sensor may be inclined sideways in a programmed mode in steps from $-27°$ to $+27°$ to cover any point on earth, subject to cloud cover permitting it, in five days. Three operating Spot satellites can do so in one day.

Digital electro-optical cameras have also been constructed by the DLR for use on Mars. After the Russian spacecraft failed on launch, the DLR constructed an aircraft version of that camera. Its design is now being manufactured by the Leica-Helava Systems Company for aerial surveys as the ADS 40 airborne digital sensor (www.lh-systems.com).

A raw and a rectified image of that sensor are shown in Figures 2.32 and 2.33.

The digital sensor of the Zeiss-Intergraph Imaging Company, the digital modular camera (DMC), has split objectives imaging onto matrix sensors of seven image planes.

Table 2.7 Parameters of the Spot electro-optical scanner

Name	ω	Ω	No. of channels	Type
SPOT-P	0.012 mrad	4.2°	1	panchromatic
SPOT.XS	0.024 mrad	4.2°	3	multispectral

Figure 2.31 The LH Systems ADS 40 digital camera.

Source: Image courtesy of LH Systems (Leica Geosystems), San Diego, CA. © Leica Geosystems, 2000.

Figure 2.32 Raw ADS 40 image.

Source: Image courtesy of LH Systems (Leica Geosystems), San Diego, CA. © Leica Geosystems, 2000.

Figure 2.33 Rectified ADS 40 image.

Source: Image courtesy of LH Systems (Leica Geosystems), San Diego, CA. © Leica Geosystems, 2000.

Figure 2.34 The Z/I DMC digital camera.

Source: Image courtesy of Z/I Imaging Corp., Huntsville, AL.

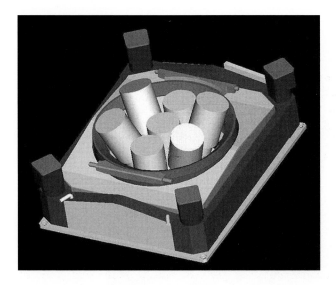

Figure 2.35 The Z/I DMC split objectives.

Source: Image courtesy of Z/I Imaging Corp., Huntsville, AL.

The images are resampled and reconstructed in a single frame by software (www.ziimaging.com).

The realization of a stereo-electro-optical scanner is due to O. Hofmann at MBB. It was built for use in the Space Shuttle and MIR for the MOMS 02 and MOMS 02-P satellite missions (see Figure 2.36) (http://isis.dlr.de).

The MOMS 02 sensor possesses a vertical CCD line with 5 m to 6 m ground pixel and two forward- and aft-looking CCD arrays with 15 m to 18 m pixels in the same image plane in the panchromatic range. Parallel to the vertically looking panchromatic CCD, there are three additional filtered multispectral arrays at 15 m to 18 m ground pixels.

For stereo sensing, only the vertical panchromatic and the two forward- and aft-looking CCD lines are used. Further developments of this principle have been introduced in the Spot 5 satellite of 2002 yielding a ground pixel of 2.5 m, paired with an equivalent array oriented at a fixed angle.

Image spectrometers

It is possible to design an opto-electronic scanner with an opto-electronic array of photosensitive elements in conjunction with diffractive gratings in such a manner that, in combination with a continuous variable optical filter, narrow bands of only 10 mm wavelength are projected onto the array resulting in continuous spectral signatures in the visible and infrared spectrum. These so-called hyperspectral devices have the ability to image

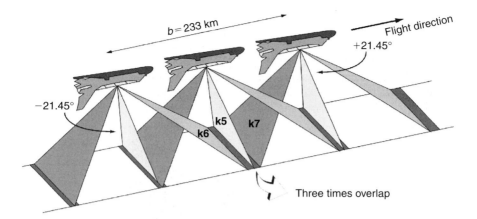

Figure 2.36 MOMS stereo scanner.

in up to 224 spectral channels, which can be compared with object libraries. The spatial resolution of these hyperspectral devices has, of course, to be reduced in accordance with the reflected energy available. For the AVNIR, visible and near infrared sensor imaging sixty bands in 10 mm increments between 430 nm and 1012 nm wavelength, operating at an altitude of 1600 m, a pixel size of 0.8 m can be reached. The AVIRIS spectrometer of NASA-JPL is operated from high-altitude aircraft ($h = 20$ km). It has 224 spectral channels at intervals of 10 nm between 400 nm and 2450 nm wavelength. Its ground pixel size at that altitude is 17 m and the swath 11 km (www.aviris.jpl.nasa.gov).

NASA has launched a TRW-built 200 channel image spectrometer Hyperion on the EO-1 spacecraft in the year 2000, with 30 m ground pixels at a swath of 7.5 km. A comprehensive list of imaging spectrometers is given in www.geo.unizh.ch~shaep/research/apex/is_list.htn.

Radar imaging

The natural radiation in the microwave range of the electromagnetic spectrum is generally too weak to be useful for imaging. Therefore, passive sensing is rare. Radar imaging, therefore, utilizes an active sensor, generating the transmitted and reflected energy in the microwave region.

Radar systems have principally been built in three wavelength regions:

- X band, $\lambda = 2.4$ cm to 3.8 cm (8000 to 12 500 MHz).
- C band, $\lambda = 3.8$ cm to 7.5 cm (4000 to 8000 MHz).
- L band, $\lambda = 15$ cm to 30 cm (1000 to 2000 MHz).

The X band and the C band in particular have the advantage of cloud penetration. They can therefore operate day and night in an all-weather system.

Let us now look at the simple operating principle of a radar. A transmitter generates a radar pulse composed of a wave signal in the respective band. The duration of the radar pulse is Δt. It is transmitted through an antenna with special propagation characteristics, so that the energy is concentrated in a narrow beam perpendicular to the platform motion. It reaches the ground with the speed of propagation of electromagnetic waves v

$$v = \frac{c}{n},$$

c being the velocity in vacuum and n the propagation coefficient.

At the terrain, the energy is directionally reflected, scattered or absorbed. The reflected energy retro-reflected into the direction of the transmitting antenna is characterized by the radar equation for the radiant flux received:

$$\phi = \frac{\phi_o \cdot G^2 \cdot \lambda^2 \cdot \rho_o \cdot A}{(4\pi)^3 \cdot r^4} \cdot 10^{-0.2 \cdot \alpha \cdot r}$$

with the following quantities:

ϕ_o = transmitted radial flux from the antenna in W.
G = antenna gain along the direction of transmission.
ρ_o = retro-reflection or backscattering coefficient of the terrain point.
A = reflecting surface in m^2.
r_i = distance between antenna and object i.
α = coefficient of atmospheric attenuation, which is wavelength λ dependent.

The transmitted and backscattered radar pulse of different terrain points along the plane of transmission reaches the antenna at different times, T_i:

$$T_i = \frac{2r_i}{v}$$

The achievable ground resolution of a radar system in the direction perpendicular to the platform motion, a, depends on the duration of the pulse, Δt, and the depression angle, β, between horizon and the transmitted and reflected ray:

$$a = \frac{v \cdot \Delta t}{2 \cdot \cos \beta}$$

For the reception of the backscattered energy, the antenna is switched by a duplexer from transmission to reception. This permits the recording of the incoming signals as a function of T_i. When the reception from all the terrain points in the plane is completed, the antenna is again switched to the transmitter and a new pulse is sent while the platform has moved forward.

The resolution in the direction of forward motion, b, like for scanners, depends on the platform velocity, v_g, and the time interval ΔT, between successive pulse transmissions

$$b = v_g \cdot \Delta T$$

However, since it is difficult to bundle the transmitted energy in one plane, the time interval, ΔT, at which two successive pulses may be transmitted, equally depends on the antenna characteristics. The azimuthal dimension $\Theta°$ of a radar beam in the transmission plane depends on the length, L, in metres of the transmitting antenna, for which the following relation is valid:

$$\Theta° = 60 \cdot \frac{\lambda}{L} \text{ and } b = r \cdot \Theta° = \frac{b}{\sin \beta} \cdot \Theta°$$

This limits the azimuthal resolution of a side-looking airborne radar (SLAR), since the length of an antenna depends on the length of a platform (e.g. an aeroplane). Radar imaging is illustrated in Figure 2.37.

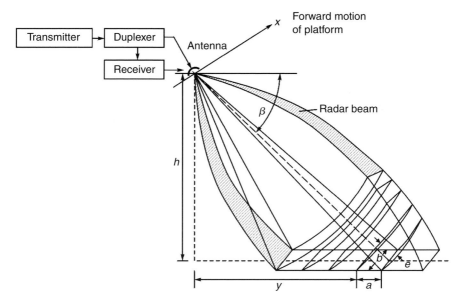

Figure 2.37 Radar imaging.

There is a better possibility to improve the azimuthal resolution with small antennas, which radiate the energy in a wide beam. Since the radar pulse emits a wave at a known frequency, the coherent energy from the reflected target not only permits the uses to determine its intensity, but also to use its frequency information. A target's position along the flight determines its Doppler frequency of its backscatter. Targets ahead of the aircraft produce a positive Doppler offset, those behind the aircraft produce a negative Doppler frequency.

Thus, the signal may be geometrically focused to a Doppler frequency of zero. This is analogous to a holographic reconstruction of the wave signals to form an image. The image coordinate in azimuthal direction can be generated as a slant range distance, y'_s:

$$y'_s = \frac{v \cdot T}{2} \cdot m_y = m_y \cdot r_s$$

in which v is the velocity of wave propagation, T is the time interval between emission and reception and m_y is a scale factor. r_s is the slant range.

The slant range distance y'_s can be reduced to a ground range distance, y'_G, for a specified platform elevation, h:

$$y'_G = m_y \cdot \sqrt{r_s^2 - h^2}$$

The image coordinate in flight direction is:

$$x' = m_x \cdot x$$

The scale factor, m_x, is a function of the platform velocity, v_g.

Height differences of the terrain, Δh, cause slant range differences, Δr, or their horizontal projection, Δy. According to Figure 2.38:

$$\Delta r = \sqrt{y^2 + (h - \Delta h)^2}$$

or

$$\Delta y = \sqrt{y^2 - 2\Delta h h + \Delta h^2}$$

Polarization of a radar beam

Most side-looking airborne radar (SLAR) or synthetic aperture radar (SAR) systems have antennas emitting the radar pulses in a polarization plane. Most frequently the horizontal polarization, H, is chosen, but it is also possible to use a vertical polarization, V.

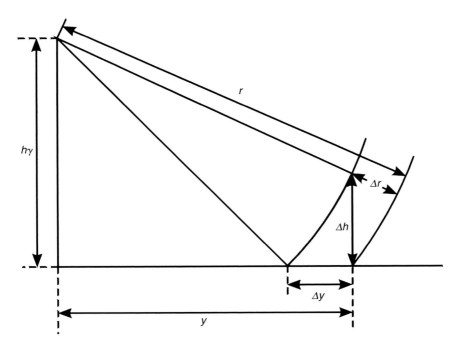

Figure 2.38 Height displacements in the radar image.

Due to the fact that radar backscattering objects may depolarize the reflected signal in all directions, it becomes possible to receive the radar backscatter in either the horizontal or the vertical polarization plane. The following combinations are therefore possible: *HH, HV, VV* or *VH*, of which generally only three polarization returns are independent, since $HV \approx HV$. Multipolarization therefore adds another dimension to radar remote sensing.

Radar interferometry

Radar pulses are transmitted as coherent waves, and they are reflected as such. If the returning waves are received by two spatially separated antennas, then the two wave signals may be compared with respect to the phase difference by means of interferometry.

The interferometric principle is shown in Figure 2.39.

The axis of the interferometer perpendicular to the base, *b*, yields signal differences with a phase difference of zero. At an angle to that axis, phase differences will be observed which are proportional to an angle in relation to the base and half the wavelength. This permits the location of the surface in terms of a digital elevation model.

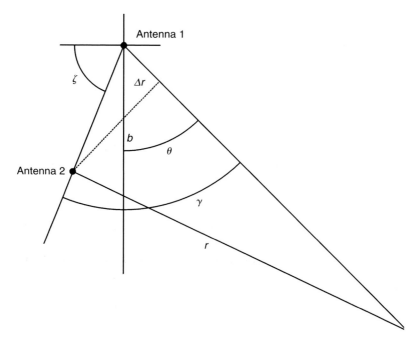

Figure 2.39 Radar interferometry.

If the phase arriving at antenna 1 (transmission and reception) is:

$$\phi_1 = -\frac{4\pi}{\lambda} \cdot r$$

and the phase arriving at antenna 2 (reception only) is:

$$\phi_2 = -\frac{4\pi}{\lambda}(r + \Delta r)$$

then the phase difference to be observed is:

$$\Delta\phi = \phi_1 - \phi_2 = \frac{4\pi}{\lambda}\Delta r$$

Introducing the base, b, between the antennas, and the angles, ξ, between horizon and base as well as the angle, θ, between the point of the reflection and the vertical, one obtains

$$(r + \Delta r)^2 = r^2 + b^2 - 2rb\cos\gamma$$

with $\gamma = 90° - \xi + \theta$
  Thus:

$$\Delta r = \sqrt{r^2 + b^2 + 2rb\sin(\theta - \xi)} - r^2 \approx b\sin(\theta - \xi)$$

The distance, y, from the flight axis, where a phase difference of 0 occurs is:

$$y = r\sin\theta$$

The next 0 phase difference occurs at a location:

$$y = r\sin(\theta - \xi)$$

The phase differences can be made visible in the form of interferometric fringes.

Figure 2.40 shows a radar image, Plate 1, colour section, shows the generated interferometric fringes and Figure 2.41 the derived digital elevation model.

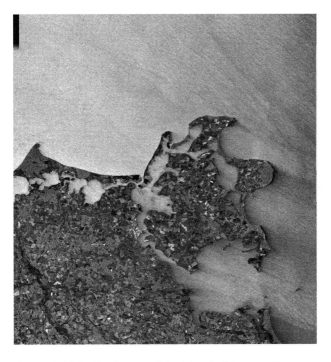

Figure 2.40 Radar image of the island of Rügen, Germany.
Source: ERS-1 SAR © ESA, processed by DLR, courtesy of DLR, Oberpfaffenhofen.

Figure 2.41 Digital elevation model derived from interferogram for the area of Hannover, Germany.

Source: DEM from ERS 1/2 SAR © ESA, courtesy of DLR, Oberpfaffenhofen.

For airborne radars, a second antenna may easily be accommodated on the aircraft. For satellites this is more difficult. During the Shuttle Radar Topographic Mission (SRTM) flown by NASA and the German DLR on the space shuttle, the second reception antenna was placed on a 60 m-long beam extended from the shuttle during the radar-mapping mission. The spatial position of the two antennas and the length of the base were determined by GPS receivers.

It is also possible to create interferograms from radar signals of two different satellites, as was done during the European Space Agency's ERS1/ERS2 Tandem Mission, in which the second satellite, ERS2, followed the first, ERS1, one day later in the same orbit. On the assumption that orbital differences created a small base of between 100 m to 300 m, and that the radar reflection properties have not changed during one day, interferometry became possible. ERS1 and ERS2, however, did not have precise orbital data, so that the length and the orientation of the base had to be estimated. This made the interferometric fringes ambiguous, and a restitution of the interferogram into digital elevation models required a trial and error 'phase unwrapping' procedure.

For interferograms produced from the ERS 1/2 Tandem Mission, the agreement with a precise digital elevation model was within 5 m in open flat areas, although it was considerably less in forested areas, and it reached 100 m in mountain areas due to radar shadows and foreshortening.

Platforms

Aircraft

The classic sensor platform is the aircraft. In order to systematically cover a portion of the earth's surface by aerial photography, flight planning is required. The aerial flights are arranged in parallel strips allowing a sufficient overlap of imaged areas by about 20 per cent to 30 per cent. Along the flight axis, an overlap of 60 per cent is generally chosen, so that an overlapping pair of photos may permit the location of any photo point in at least two photographs. Since two photos are required from different exposure stations to determine an object point in three dimensions the overlap scheme not only fulfils this condition, but also permits the construction of a strong geometric interconnection between the adjacent images.

The flight-planning scheme is shown in Figure 2.42.

An aerial photograph has the standardized dimension, a', of 23 cm. Thus the area covered by a vertical photograph $A = a^2$, in which

$$a = \frac{h}{f} a'$$

with the flying height, h, and the principal distance, f.

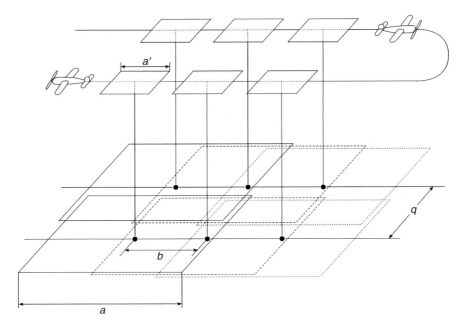

Figure 2.42 Flight plan.

The base between two subsequent photos in the strip becomes

$$b = a\left(1 - \frac{o}{100}\right)$$

in which o is the chosen longitudinal overlap of 60 per cent. Thus b is usually

$$b = a \cdot 0.4$$

When the overlap between strips, p, is 20 per cent, then the distance between two strips, q, becomes:

$$q = a\left(1 - \frac{p}{100}\right) = a \cdot 0.8$$

A single model composed of the interior parts of the two photographs is of significance to calculate the number of photos required. This neat area, N, has the dimensions $N = b \cdot q$.

Thus, the number of photos, n, required to cover a total area, B, is:

$$n = \frac{B}{N}$$

For these photos, a film length of $0.25\,\text{m} \times n$ will be required. The scale of the image is 1:scale factor. The image scale factor is given by the ratio h/f. h is the flying height above ground. To reach the required overlap conditions, this height is always taken at the maximum altitudes of the terrain to be imaged.

Among the possible lens cones to be chosen for the flight are wide-angle objectives with the principal distance of 153 mm or normal angle objectives with the principal distance of 305 mm. In special cases, super wide-angle objectives with a principal distance of 85 mm to 88 mm may be utilized.

Wide-angle objectives are the most widely used. They offer about equal accuracy to determine positions and elevations. However, they have larger displacements by elevated objects such as buildings and trees. Thus, larger areas may become hidden through these objects. For this reason, urban and forest surveys in general prefer normal angle photography.

The aircraft altitude depends on the nature of the aircraft. Only military planes reach altitudes between 12 km and 25 km. From 8 km to 12 km, jet aircraft are required. Below 8 km, turbo prop and propeller-driven aeroplanes may be used. The lowest altitude, h_{min}, at which aerial flights can be made depends on the flight velocity with respect to the ground, v, and the required minimum time interval between exposures fixed by the camera, Δt_{min}:

$$b_{min} = \frac{v}{a'\left(1 - \dfrac{0}{100}\right)} \cdot \Delta t_{min}$$

As a rule, aerial flights are only made in clear, cloudless weather conditions when the solar altitude is higher than 30 per cent, but it should not exceed 60°. This limits the flying season for certain areas of the globe. For topographic surveys, a winter flying season is preferred, when the foliage of trees is minimal. Central Europe, in general, only has about 22 to 28 days per year for which aerial surveys can be made.

Aerial flight navigation before the advent of GPS required the use of a variety of electronic navigation systems. Today automatic GPS navigation with devices offered by the camera manufacturers and their affiliates have become the rule (CCNS by IGI, T-Flight and POS Z/I by Z/I Imaging, Ascot by LH Systems).

Two companies, Applanix of Canada and IGI of Germany, offer additional inertial devices, which permit the recording of flight attitudes in three axes of rotation, in addition to determining the coordinates of the exposure station at the time of exposure.

While relative positioning via DGPS is possible to about ±10 cm to 15 cm, angular parameters of ±0.003° may be obtained for pitch (φ) and roll (ω) and of ±0.007° for yaw (κ) by inertial devices. This is achieved by accelerometers for which the signals are integrated in an inertial measuring unit (IMU). Boresight calibration procedures are required before the flight to resolve the transformation between the three spatial coordinate systems for IMU, GPS and camera.

While the purchase of an IMU is still very expensive, it has been shown that satisfactory operations are possible to cover large areas in high altitude, small-scale flights (<1:30 000) without the use of ground control and the need for an aerotriangulation. This is particularly useful for large orthophoto projects. Also at larger photographic scales, GPS positioning and IMU attitude data can be input into aerial triangulation block adjustments to minimize the required ground control.

Satellites

After the launch of the first satellite Sputnik by the former USSR in 1957, the first US satellites such as Tiros 1 in 1960 began to carry remote sensing devices to image weather patterns.

An undisturbed satellite can stay in circular orbit, if its velocity, v, is chosen in accordance with the mass, M, around which the satellite orbits, the gravitational constant, G, and the radius of the orbit from the centre of mass, r:

$$v = \sqrt{\frac{G \cdot M}{r}}$$

GM for the earth is a constant $= 3.980 \cdot 10^5 \, km^3/sec^2$,

$$r = r_o + h$$

with $r_o = 6370 \, km$ and h is the satellite altitude above the surface.

The period of one revolution, U, in minutes of time, as derived from Kepler's third law, is:

$$U = 84.491 \cdot \sqrt{\frac{r^3}{r^3_o}}$$

This means that a great number of satellites can be kept in orbit, as shown in Table 2.8.

Most earth-orbiting satellites have near circular orbits for which these simplified relations are valid.

Another orbital characteristic is the inclination, I, expressed as an angle between the orbital plane and the equatorial plane, and its relation to the vernal equinox.

Geostationary satellites orbit in the equatorial plane of the earth at a speed equivalent to the earth's rotation. These satellite orbits are ideal for communication satellites and for global weather satellites looking at the entire hemisphere (Meteosat, GOES1 and 2, GMS and Insat). Their longitude positions are at 0°, 75°W, 135°W, 140°E and 74°E.

Most earth observation satellites (Landsat, Spot, etc.) prefer imaging in the mode of a sun-synchronous satellite. This is possible when a constant relation between orbital node and the direction to the sun is maintained.

The orbit with its inclination and the earth rotation determine the

Table 2.8 Satellite characteristics

r	h	v	U	Remarks
6700 km	330 km	7.71 km/s	90.97 min	space station
7370 km	1000 km	7.34 km/s	105.6 min	earth observation satellite
26 570 km	20 200 km	3.87 km/s	12 h	GPS
42 160 km	35 790 km	3.07 km/s	23 h 56 m	geostationary satellite
384 400 km	–	1.02 km/s	27 d 08 m	moon

ground track of the satellite. Repetition of that track after a specified number of days is advantageous for earth observation satellites, permitting the gathering of images within a predetermined pattern.

A satellite is subjected to various orbit-degrading influences (e.g. solar drag). To keep it in the predetermined orbit, fuel must be carried on board. It is used to make occasional thrust manoeuvres to achieve the required orbit corrections.

Ground resolution versus repeatability

Figure 2.43 illustrates the choice of a remote sensing system with its achievable resolution and with the repeatability to obtain data.

Geostationary meteorological satellites such as Meteosat and GOES permit the imaging of the entire hemisphere, as seen from an altitude of 35 790 km, by an electro-mechanical scanning radiometer kept in constant

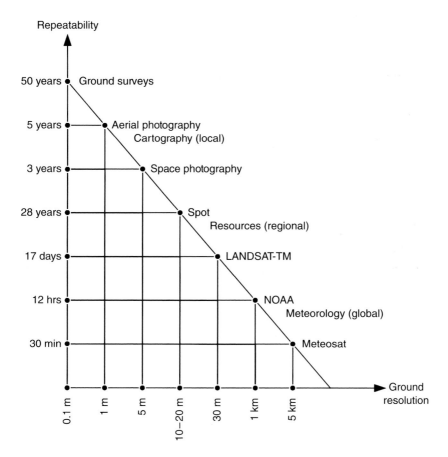

Figure 2.43 Ground resolution versus repeatability.

rotation with 5 km pixels in three bands, one of which is thermal. This can be achieved every 30 minutes. Meteosat images are placed on the Internet at least three times a day.

Geostationary satellites permit the study of weather patterns throughout the day, viewing moving clouds on TV, but they are too coarse in resolution for vegetation studies and for viewing the polar regions. Thus NOAA satellites with a polar orbit and an altitude of about 850 km permit the imaging of the earth every 12 hours at 1 m pixels at a swath of 2700 km in five channels.

Sunsynchronous remote sensing satellites such as Landsat (USA), Spot (France) and IRS (India) have a repetition rate of less than a month but, due to high cloud cover probability, an image may be obtained only several times per year at medium resolutions between 5 m and 15 m in panchromatic and between 20 m and 30 m in multispectral mode at swaths between 185 km to 60 km.

High-resolution systems from space, such as Ikonos 2 with 1 m pixel and Russian space photography at 2 m resolution attempt to compete with high-altitude aerial photography. However, a high area coverage with repetitivity of less than every few years is not possible.

The highest accuracy in the centimetre to decimetre range is achieved by low-altitude aerial surveys and by ground surveys. Whether these methods can be utilized over large areas is dependent on questions of cost and time. It is appropriate to apply low-resolution, high-repetitivity sensors for global surveys, intermediate resolution and intermediate repetitivity surveys, and high-resolution and low-repetitivity sensors for local (e.g. urban) surveys.

Image interpretation

The image generated by a remote sensing sensor is subject to interpretation, before the remote sensing data can become information. While there is research looking at an automation process for the information extraction procedure, currently all practical interpretations are based on the human eye–brain system.

The human eye

The eye performs the task of optical imaging, while the brain performs the analysis of the perceived optical data. Figure 2.44 describes the composition of the human eye.

The eye possesses a lens, which can change its curvature for focusing a near or far object onto the retina, creating an image. This change of focus is achieved by the movement of a muscle. The incoming light intensity on the retina is regulated by the pigmented part of the eye controlling the variable aperture (the pupil). The part of the retina with the highest con-

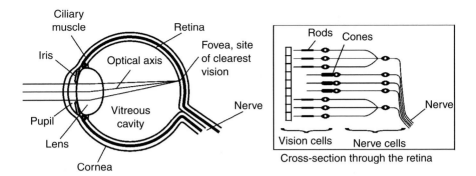

Figure 2.44 The human eye.

centration of light-sensitive cells is the fovea, extending 1° to 2° from the optical axis of the eye, while peripheral, low-resolution vision is possible in a range of 120°.

The young eye can change its focal length between 17 mm and 23 mm. The attainable resolution in the fovea is about 7 lp/mm. An aerial photo of 40 lp/mm can thus be observed at 5- to 6-fold magnification by a lens system.

The light-sensitive elements are the more densely spaced rods suitable for panchromatic sensing at low energy levels. The more widely spaced cones permit the observation of spectrally mixed energy in the form of colour. The eye contains about 10^8 rods and 10^7 cones.

The lens system of the eye projects an inverted image onto the retina. The erect perception of the image is a function of the brain: the rods and cones cause chemically produced signals transmitted to the visual area of the cortex of the brain for signal processing by neural networks.

This processing analyses the images received for grey level, colour, texture, size and context and motion, and converts them into information through a comparison with stored information in the neurons (nerve cells) of the brain. With about 10^{11} neurons contained in the brain, visual perception by far exceeds the image analysis capabilities of a computer.

Stereo vision

Another interpretation tool exists in human vision; two eyes permit the fusion of two images taken from spatially different observation points, allowing a judgement of the distance of the observed object, y. Figure 2.45 shows the capacity for natural stereoscopic vision.

The human eye base, b_E, is about 65 mm. It approximately determines

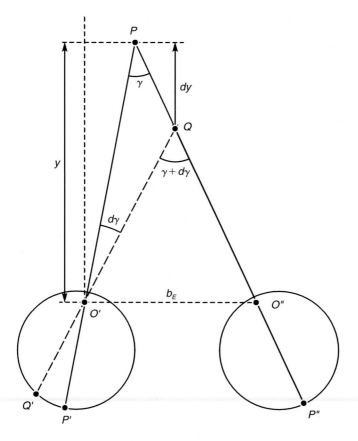

Figure 2.45 Natural stereoscopic vision.

γ, which is a small angle

$$\gamma \approx \frac{b_E}{y}.$$

By differentiation:

$$d\gamma = -\frac{b_E}{y^2} dy$$

stating that natural stereoscopic vision diminishes with the distance squared. The stereoscopic observation capability of the brain is $d\gamma = 15''$. Thus the ability of the brain to judge distance differences, dy, at a distance, y, is listed in Table 2.9.

Table 2.9 Natural stereoscopic depth perception limit

y	0.25 m	10 m	100 m	500 m	894 m
dy	0.07 mm	0.1 m	11 m	280 m	894 m

The ability to fuse images is furthermore limited by the angular range between farthest object, γ_F, and closed object, γ_c:

$$\gamma_F - \gamma_c \leq 70'$$

Image interpretation and photogrammetry have the possibility to expand the stereoscopic observation capacity to judge and to measure distances stereoscopically through the use of images, which have been taken at n-times magnification of the eye base. If the two images are presented to both eyes with a magnification, m, and at a magnification of the eye base

$$n = \frac{b}{b_E}$$

then:

$$dy = -\frac{1}{n \cdot m} \cdot \frac{y^2}{b_E} \cdot d\gamma = -\frac{1}{m} \cdot \frac{y^2}{b} \cdot d\gamma$$

Two aerial photographs taken from a distance (flying height) of 1000 m, viewed at a four-times enlargement and at 60 per cent overlap with a base, b, of 400 m thus permit a stereoscopic height determination from the images at 0.05 m.

For the observation of stereo adjacent aerial photographs, it is necessary to orient the images according to epipolar rays (see Figure 2.46).

Epipolar rays are the lines of projection with the plane formed by the object point and the projection centres of the two images (see Figure 2.47).

In practice, it is helpful to transfer the principal points H'_1 from left to right H''_1 and from right H''_2 to left image H'_2 and to position the images along a straight line containing H'_1, H'_2, H''_1 and H''_2.

Along this line (and parallel to it) the image points L′M′N′ and L″M″N″ can be viewed in stereo as points L, M, N.

The brain is able to compensate for minor differences in that direction of about 2 per cent, as well as scale differences up to 5 per cent.

The images are easiest observed with lens stereoscopes having an eye base of 65 mm and a lens magnification of 1.6. The images placed along the epipolar line then require a separation of 65 mm (see Figure 2.48).

Aerial photographs of the size 23×23 m are better observed in a mirror

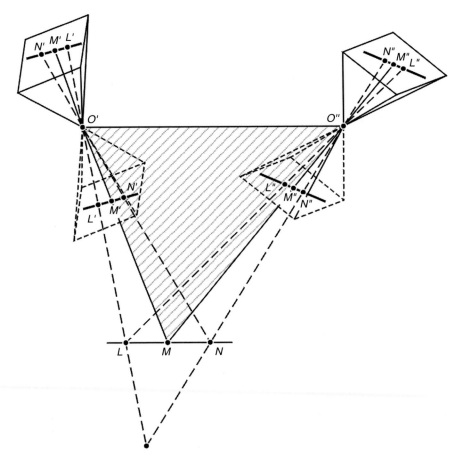

Figure 2.46 Epipolar plane and epipolar rays.

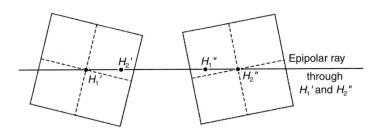

Figure 2.47 Orientation of aerial photographs according to epipolar rays.

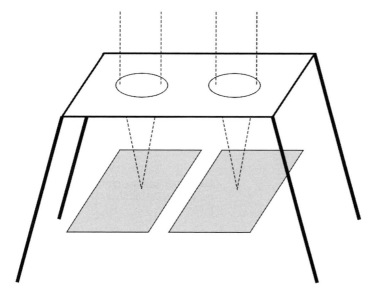

Figure 2.48 Lens stereoscope.

stereoscope which separates the images by a mirror system and allows a magnification of up to four times (see Figure 2.49).

Stereo observation is also possible via anaglyphs in complementary colours (red and green) when viewed through corresponding filters. The anaglyphic images are projected or printed on top of each other in the respective colours, and viewing through filtered spectacles is possible without lenses (see Plate 2, colour section).

Colour images may be viewed if they are projected in two polarizations

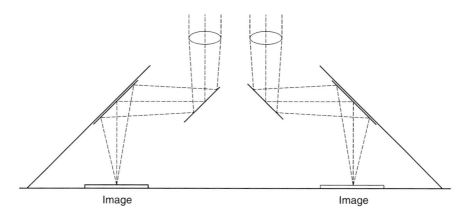

Image Image

Figure 2.49 Mirror stereoscope.

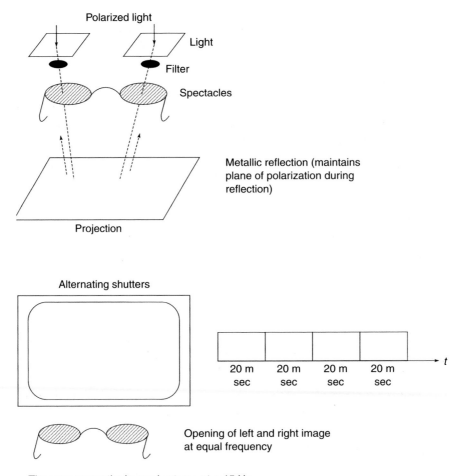

Figure 2.50 Stereo viewing by polarized light and alternating shutters.

and viewed with corresponding polarization filters. To obtain a polarized image, not only the projected light onto a projection plane needs to be polarized, but also the projection surface needs to reflect the projected light in polarized mode. This is possible for metallic projection surfaces (see Figure 2.50).

Stereo viewing on a computer screen becomes possible by split screens when viewed through a stereoscope. Anaglyphs may also be viewed on the screen. Special screens permit viewing with polarized light. As a rule, the 'Crystal Eyes' principle is used, in which the left and right images are alternately generated at a 50 Hz rate. These can be viewed with filtered spectacles, which open and close the alternating left and right view at the same 50 Hz frequency.

Stereovision greatly assists in the interpretation possibilities of objects. It is also the basis for the manual photogrammetric restitution process.

Visual interpretation of images

Based upon the possibilities given by the human eye–brain system, the interpretation of images by an analyst starts at a primary level observing contrasts of tone and colour. At a secondary level, size, shape and texture are compared. At the third level, pattern, height difference and shadow aids in the interpretation. At a fourth level, the association with adjacent objects plays a role.

Image interpretation has a great number of areas of application. These include:

- military intelligence,
- forestry,
- agriculture,
- hydrology,
- topographic mapping,
- urban analysis,
- coastal area surveys,
- and archaeology, among others.

For some of these applications, interpretation keys with examples of imaged objects to be interpreted have been developed.

Image processing

Raster scanning of photographic images

If the remotely sensed images have not been obtained by digital sensor, but by photographic imaging, the application of automated computer analysis requires a raster digitization of these images. One of the first widely used devices for this purpose, from around 1970, was the Optronics Scanner, shown in Figure 2.51.

It consisted of a rotating drum, onto which the film transparency was wrapped. While the image rotated, an electronically controlled lamp sent a flash of light through the aperture and the emulsion, and collected the energy on a photo multiplier. This analogue signal was amplified and digitally converted into grey levels for each exposed pixel. After exposure of an image line, the aperture was shifted by a step motor and the next line was digitized in the same manner. If desired, the scanning of the image could easily be combined with a photo-write mechanism activated by a

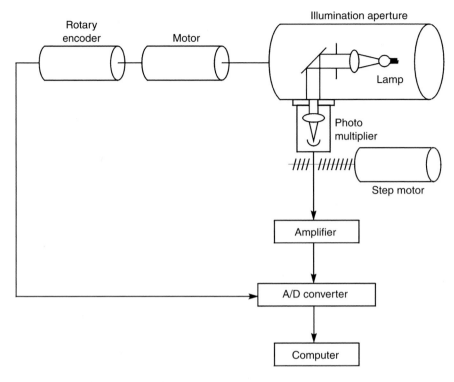

Figure 2.51 The Optronics drum scanner.

laser diode. This permitted online image processing under control of a computer but, as a rule, it was sufficient to store the image on a suitable storage medium. Colour images could be scanned sequentially using three-colour filters only permitting the recording of blue, green and red light.

The Optronics permitted digitization in pixel increments of 12.5 μm, 25 μm, 50 μm and 100 μm. Mechanical and thermal instabilities of the instrument, however, limited the practical use to only 50 μm pixels. To reach higher digitizing resolutions, flat-bed scanners have been introduced. Table 2.10 gives a summary of the available devices. Figure 2.52 demonstrates the principle of a flat-bed scanner.

Figure 2.53 shows the Z/I Imaging Photoscan 2001.

Grey level changes

Grey level changes are single pixel-based operations, which permit the changing of the available digital density of a pixel, d_i, by changing the

Table 2.10 Flat-bed scanners for aerial photographs

Manufacturer	Name	Sensor	Colour	Image pixel size	Radiometric range	Film
Leica	DSW 500	camera with 2048 × 2048 pixel patches, stair stepped trilinear array	yes	continuous 4 to 20 μm	0.1–2.5 D	single or roll film
Vexcel	Ultrascan 5000		yes	continuous 2.5 to 2500 μm	0–4 D	single or roll film
Wehrli	RM2 Rastermaster	linear array	no	10 μm	0.1–2 D	single film
Z/I	Photoscan 2001	trilinear CCD	yes	7, 14, 21, 28, 56, 112, 224 μm	0.1–3 D	single or roll film

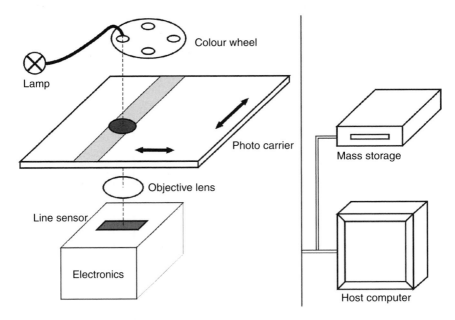

Figure 2.52 Principle of a flat-bed scanner.

Figure 2.53 Z/I Photoscan 2001.

Source: Image courtesy of Z/I Imaging Corp. (Photoscan tm), Huntsville, AL.

analogue D-logH or the digital D-H curve into a new pixel density d'_i. Examples of the possibilities are shown in Figure 2.54:

- for a change of the slope of the γ curve or the response curve:

$$d'_i = a_o + a_1 d_i$$

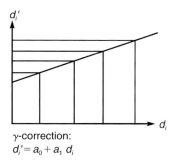

γ-correction:
$d_i' = a_0 + a_1\, d_i$

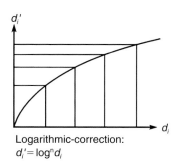

Logarithmic-correction:
$d_i' = \log^n d_i$

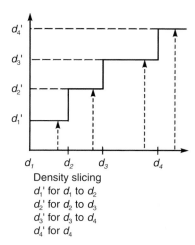

Density slicing
d_1' for d_1 to d_2
d_2' for d_2 to d_3
d_3' for d_3 to d_4
d_4' for d_4

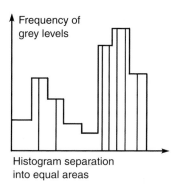

Histogram separation
into equal areas

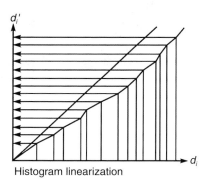

Histogram linearization

Figure 2.54 Grey level changes.

- the response may also be logarithmically changed:

  $d'_i = \log^n d_i$ or

- densities may be grouped into preselected density ranges, by a step function.
- finally, a grey level adaptation may be achieved by a histogram linearization, in which the surface under the histogram is divided into equal areas, defining new limits, d_i, which are to be imaged as d'_i with constant intervals. This is shown in Figure 2.54.

Grey level changes permit the display of an image at better contrast conditions for viewing, since the eye favours a density range between 0.1 and 1.0 d.

Filtering

Filtering operations involve the adjacent pixels for each pixel. Local filters operate on an image matrix, $D(x, y)$, with a finite, usually small dimension, e.g. (3×3), (5×5), etc. For this matrix there exists an assigned weight matrix, $W(x, y)$. The result of the filtering process for each pixel, $d(x, y)$, results out of a convolution of the image matrix, $D(x, y)$, with the weight matrix, $W(x, y)$.

Examples for a 3×3 image matrix, $D(x, y)$, and a number of important weight matrices, $W(x, y)$, are shown in Figures 2.55 and 2.56.

The result of the convolution for pixel, i, j, is:

$$D'(i, j) = W(x, y) \, ^* D(x, y)$$

- Applied to a low pass filter this means:

$$d'_{ij} = \frac{1}{9} \left(\sum_{k=i-1}^{k=i+1} \sum_{\ell=j-1}^{\ell=j+1} d_{k,\ell} \right)$$

  low pass.
- Applied to a vertical directional contrast:

$$d'_{ij} = \sum_{k=i-1}^{k=i+1} |d_{k,j-1} - d_{k,j+1}|$$

  vertical contrast.
- Applied to a horizontal directional contrast:

$$d'_{ij} = \sum_{k=j-1}^{k=j+1} |d_{i-1,k} - d_{i+1,k}|$$

  horizontal contrast.

Image $D(x, y)$

$d_{i-1, j-1}$	$d_{i, j-1}$	$d_{i+1, j-1}$
$d_{i-1, j}$	$d_{i, j}$	$d_{i+1, j}$
$d_{i-1, j+1}$	$d_{i, j+1}$	$d_{i+1, j+1}$

Filter $W(x, y)$

1/9	1/9	1/9
1/9	1/9	1/9
1/9	1/9	1/9

Low pass

+1/3	+1/3	+1/3
0	0	0
−1/3	−1/3	−1/3

Vertical directional

+1/3	0	−1/3
+1/3	0	−1/3
+1/3	0	−1/3

Horizontal directional

0	+1	+1
−1	0	+1
−1	−1	0

Diagonal directional

+1	+1	0
+1	0	−1
0	−1	−1

Diagonal directional

Figure 2.55 Low pass filter and directional filters.

Diagonal contrasts can be obtained in an analogous manner by the shown filters.

- The Laplace operator for edge enhancement is:

$$\Delta d(i, j) = \frac{\delta^2 d_{ij}}{\delta x^2} + \frac{\delta^2 d_{ij}}{\delta y^2}$$

0	−1	0
−1	4	−1
0	−1	0

Laplace
edge enhancement

1	2	1
0	0	0
−1	−2	−1

−1	0	1
−1	0	2
−1	0	1

Sobel edge enhancement $W_1(x, y)$

Sobel edge enhancement $W_2(x, y)$

$$W = \sqrt{W_1^2 + W_2^2}$$

Figure 2.56 Edge enhancement filters.

Its approximation becomes:

$$d'_{ij} = 4d_{ij} - (d_{i-1,j} + d_{i,j-1} + d_{i+1,j} + d_{i,j+1})$$

- Edge enhancement is also possible by the Sobel operator, in which two filter matrices $W_1(x, y)$ and $W_2(x, y)$ are applied simultaneously.
- A high pass filter results from the subtraction of a low pass filtered image from the original image:

$$d'_{ij} = 2d_{ij} - d'_{i,j}$$

high   low
pass   pass

The filtering operations are executed pixel by pixel for the entire image.
Another type of filtering is possible for the entire image using the fre-

quencies at which the grey values occur. This involves calculation of a Fourier transform for the image. It is basically a coordinate transformation from the image space in Cartesian coordinates (x, y) or in polar coordinates r, α into Fourier Space (u, v), where the frequency of the signal f and α are represented as polar coordinates.

The calculation of a Fourier transform of the image, D, is:

$$FouD(u, v) = \int_{-\infty}^{+\infty} \int_{-\infty}^{+\infty} e^{2\pi\sqrt{-1}(ux+vy)} \cdot D(x, y)dx, \, dy$$

If a filter, W, is subjected to the same type of transformation:

$$FouW(u, v) = \int_{-\infty}^{+\infty} \int_{-\infty}^{+\infty} e^{2\pi\sqrt{-1}(ux+vy)} \cdot W(x, y)dx, \, dy$$

then the filtered image Fourier transform can easily be obtained by the multiplication of the two Fourier transforms in u, v space:

$$FouD'(u, v) = FouW(u, v) \cdot FouD(u, v)$$

$D'(u, v)$ can be subjected to an inverse Fourier transformation to obtain the filtered image, $D'(x, y)$, in the image space:

$$D'(x, y) = Fou^{-1}D'(u, v) = \int_{-\infty}^{+\infty} \int_{-\infty}^{+\infty} e^{-2\pi\sqrt{i}(ux+vy)} \cdot FouD'(u, v)du, \, dv$$

Use of Fourier filtering is particularly useful for the purposes of image reconstruction, when known or estimated sources of image degeneration need to be eliminated or, at least, reduced.

For the source of degradation (defocusing, image motion), a modulation transfer function $M(u, v)$ can be set up.

$$\frac{1}{M(u, v)}$$

can be used as an inverse filter $I(u, v)$.

The Fourier transform of the reconstructed image then becomes:

$$FouD'(u, v) = FouD(u, v) \cdot I(u, v)$$

in which $FouD(u, v)$ is the Fourier transform of the degraded image.

The inverse Fourier transform of $D'(u, v)$ then yields the reconstructed image $D(x, y)$, improved in sharpness:

$$D'(x, y) = Fou^{-1}D'(u, v)$$

Geometric resampling

The geometry of a two-dimensional image is distorted due to the imaging geometry of the sensor, the sensor orientation and the displacements of the three-dimensional scene, when imaged into two dimensions; for these deformations of the image models exist, which are described in Chapter 3. Such a model is a function between image coordinates, $x_i'y_i'$, and object coordinates $x_iy_iz_i$:

$$x_i' = f_1(x, y, z)$$

$$y_i' = f_2(x, y, z)$$

The inverse relations may be generated from these functions as:

$$x_i' = f_3(x', y', z)$$

$$y_i' = f_4(x', y', z)$$

With the functions f_1, f_2, f_3 and f_4 known, two resampling algorithms for geometric correction of the images may be used. They are illustrated in Figures 2.57 and 2.58.

In the direct method of resampling, the coordinate x_iy_i of an image pixel $x_i'y_i'$ is calculated with functions f_3 and f_4. The z_i is to be interpolated within the x_iy_i pixel grid. The density $d_{x'y'}'$ is transferred to that location. All calculated points are used for further interpolation of the grey values of the output pixel matrix.

In the simpler indirect resampling method, the output pixel x_iy_i coordi-

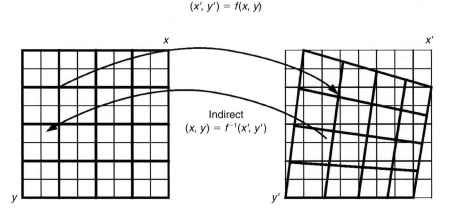

Figure 2.57 Digital rectification (1).

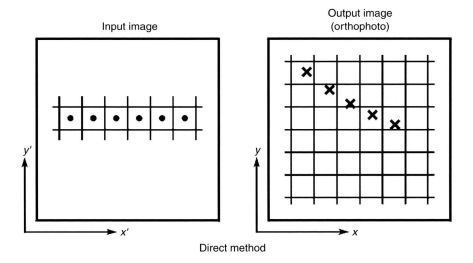

Direct method

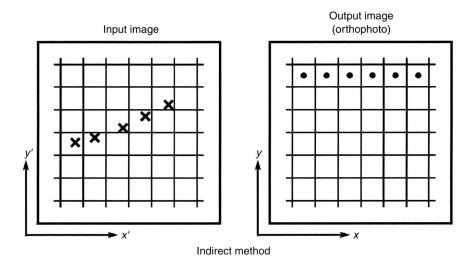

Indirect method

● Selected pixel for computation

✕ Corresponding image point after
geometric transformation

Figure 2.58 Digital rectification (2).

nates with their known or interpolated height z_i permit the calculation of the location of an image point $x'_i y'_i$ assigned to the output pixel.

For this assignment, it is possible to use three options:

- the assignment of grey values to the output pixel grid by the nearest neighbour.
- bilinear interpolation: the nearest four pixels for the calculated image point for the indirect method, or object pixels for the direct method, with densities of d_1 to d_4 and their distances to the output or input pixel centre $x_0 y_0$ are used in a weighted function:

$$d_i = \frac{p_1 d_1 + p_2 d_2 + p_3 d_3 + p_4 d_4}{p_1 + p_3 + p_3 + p_4}$$

with

$$p_k = \frac{1}{\sqrt{(x_k - x_0)^2 + (y_k - y_0)^2}}$$

with k varying from 1 to 4.
- cubic convolution: here the nearest sixteen calculated points are used in the same manner.

Multispectral classification

The objective of multispectral classification is to analyse the spectral properties of unknown objects and to compare them with spectral properties of known objects. Each spectral channel consists of a digital grey level image matrix, which geometrically coincides with the grey level image matrices of other spectral channels.

For each image a histogram of grey levels can be generated. A specific object class will produce a grey level distribution, which can be compared with a normal distribution. Statistical parameters for this comparison are, for example:

- the mean of grey levels.
- the variance or the standard deviation.
- the maximum and minimum grey levels for this object.

Two separable objects in this channel will produce two grey level distributions.

Two channels define a two-dimensional feature space, in which the grey levels of different object types are shown as clusters (see Figure 2.59).

For n channels, there exists an n-dimensional feature space. Each pixel

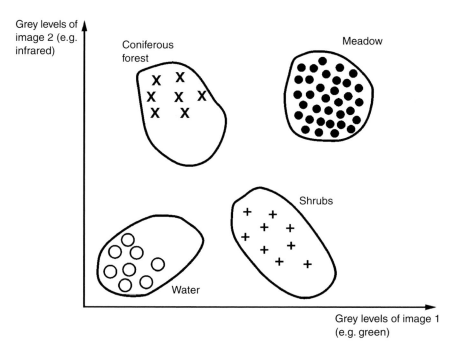

Figure 2.59 Two-dimensional feature space with object clusters.

in any of the multispectral images can be expressed by an *n*-dimensional feature vector:

$$x = \begin{pmatrix} x_1 \\ x_2 \\ \vdots \\ x_n \end{pmatrix}$$

containing the grey levels in each band.

For all object clusters, a mean value, *m*, can be formed:

$$m = \frac{1}{K} \sum_{K=1}^{K} x_K$$

The simplest type of classification is, then, for each group of pixels with the same geometry to form the minimum Euklidian distance to the cluster centres, *m*. Each pixel, *i*, obtains the classification of the closest cluster:

$$d(x_i, x_m) = \sqrt{(x_i - x_m)^T \cdot (x_i - x_m)}$$

The minimum distance classification can be refined into a maximum likelihood classification by use of covariance matrices for each cluster.

The covariance matrix, Σ_x, can be expressed as:

$$\Sigma_x = \frac{1}{K-1} \sum_{K=1}^{K} (x_K - m)(x_K - m)^T$$

From the covariance matrix, Σ_x, the correlation matrix, R, can be formed, in which the coefficients of the matrix are scaled down to a diagonal value of 1, so that the elements of the covariance matrix, v_{ij}, are transformed to:

$$r_{ij} = \frac{v_{ij}}{\sqrt{v_{ii} \cdot v_{jj}}}$$

The study of the covariance matrix or of the correlation matrix therefore permits the checking of the separability of the chosen object classes.

Each feature vector finds the probability of belonging to a certain class, ω:

$$p(x) = \frac{1}{(2\pi)^{n/2} |\Sigma|^{1/2}} \exp\left\{ -\frac{1}{2}(x-m)^T(x-m) \right\}$$

The measure of separability between two probability distributions of the classes ω_i and ω_j is the divergence, d_{ij}:

$$d_{ij} = \int_x \{ p(x|\omega_i) - p(x|\omega_j) \} \ln \frac{p(x|\omega_i)}{p(x|\omega_j)} dx$$

It represents a covariance-weighted distance between the means of two object pairs. It may be calculated as:

$$d_{ij} = \frac{1}{2} T_r \left\{ \left(\Sigma_i - \Sigma_j \right)\left(\Sigma^{-1}_i - \Sigma^{-1}_j \right) \right\} + \frac{1}{2} T_r \left\{ \left(\Sigma^{-1}_i - \Sigma^{-1}_j \right)(m_i - m_j)(m_i - m_j)^T \right\}$$

with T_r being the trace of the matrix in question.

More refined judgements are possible by the Jeffries–Matusita Distance J_{ij}, which is the distance between a pair of probability distributions:

$$J_{ij} = \int_x \{ \sqrt{p(x|\omega_i)} - \sqrt{p(x|\omega_j)} \}^2 dx$$

For normally distributed classes, it becomes the Bhattacharyya distance, B:

$$B = \frac{1}{8}(m_i - m_j)^T \left\{ \frac{\Sigma_i + \Sigma_j}{2} \right\}^{-1} (m_i - m_j) + \frac{1}{2} \ln \left\{ \frac{|(\Sigma_i + \Sigma_j)/2|}{|\Sigma_i|^{\frac{1}{2}} \cdot |\Sigma_j|^{\frac{1}{2}}} \right\}$$

The covariance matrix, Σ_x, or the correlation matrix, R, shows, that the feature vectors, x, assigned to object clusters, m, are often highly correlated between the n channels available. This permits the rotation of the feature space, x, by a rotation matrix, G, into a new feature vector space, y, so that:

$$y = Gx$$

for which the covariance matrix, Σ_y, becomes a diagonal matrix of eigenvalues, λ_i:

$$\Sigma_y = \begin{pmatrix} \lambda_1 & 0 & & 0 \\ 0 & \lambda_2 & . & 0 \\ & & \ddots & \vdots \\ 0 & 0 & \cdots & \lambda_n \end{pmatrix}$$

This transformation is called a principal component transformation. Only the equations with the largest eigenvalues suffice for an optimal separation of the chosen classes in feature space.

Classification can be performed in two different ways.

- *Supervised classification*
  It is applied if a number of object types can be recognized in the images. This implies delineation of training areas as a subset of image pixels and a generation of clusters for these areas, determining their mean vector, m. This permits the direct use of the minimum distance classifier.

  Another simple possibility is to apply a parallelpiped classifier in which the parallelpiped dimensions are formed from the maximum and minimum grey values of the training areas for a certain object, with the risk of class overlaps.

  Most appropriate is the use of the maximum likelihood classifier for each object class with its covariance matrix.

  The decision rule that x belongs to ω_i is:

  $$x \in \omega_i, \text{ if } p(\omega_i | x) > p(\omega_j | x)$$

  for all $j \neq i$.

  A somewhat simpler classification is possible by the Mahalanobis distance. In the special case, that all priority probabilities are assumed equal, the decision function becomes the Mahalanobis distance:

  $$d(x, m_i)^2 = (x - m_i)^T \Sigma^{-1} (x - m_i)$$

  The Mahalanobis classifier, like the maximum likelihood classifier, retains sensitivity to direction contrary to the minimum distance classifier.

An internal check of the classification accuracy is possible through analysis of classifications for the training areas. There it is possible to generate a confusion matrix for all object classes, listing the total numbers of pixels in each training area and their portion classified into other object classes. Obviously, an overall check in this form should be made for data obtained in the field.

* *Unsupervised classification*
  If no ground information to establish training areas is available, then clustering must be started by an iterative procedure estimating the likely location of clusters for ω objects. For example, in three-dimensional space, a set of clusters may be chosen along the diagonal at equal distances. Then a preliminary minimum distance classification is made, and the mean vector of the cluster centres is formed. Then the process is iterated.

  The obtained clusters can again be checked via the divergence to decide whether some clusters should be merged. A maximum likelihood classification can follow the process. At the end the classification result can be assigned as a plausible object class.

While the statistical approach to multispectral classification prevails in practice, another approach using neural networks is possible.

In two dimensions, a straight line may be drawn between two pixels so that:

$$w_1 x w_1 + w_2 x w_2 + w_3 = 0$$

with x_1, x_2 representing grey values and w_1, w_2, w_3 as weights. In n-dimensions, for n bands, the equation becomes:

$$w_1 x_1 + w_2 x_2 + \ldots w_n x_n + w_{n+1} = 0$$

or

$$w^T x + w_{n+1} = 0$$

For a set of image data, these weights have to be determined by a training process with the decision rules:

$$x \in class\ 1\ if\ w^T x + w_{n-1} > 0,\ or\ w^T y > 0$$

$$x \in class\ 2\ if\ w^T x + w_{n-1} < 0,\ or\ w^T y < 0$$

The weight, w, is modified to w' with a correction increment, c:

$$w' = w + cy$$

so that:

$$w'^T y = w^T y + c |y|^2$$

In practice, this is iterated until:

$$w'_i = w_i + cy$$

and

$$w'_j = w_j - cy$$

The disadvantage of the pixel based multispectral classification approach is that homogeneous objects are not treated as a unit. This can be overcome by image segmentation. Image segmentation can be implemented either by edge detection techniques or by region growing. The classification algorithms may then be applied to regions rather than to pixels. A recently developed product is the e-cognition 'context based' classifier.

If segmented data are available through a GIS system, a knowledge based classification approach may be applied to image regions. Knowledge is introduced by a set of rules: *if* a condition exists, *then* inference is applied. Figure 2.60 shows the example of a semantic network, which can be used to test segmented images for their content.

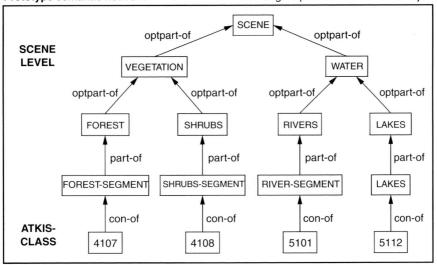

Figure 2.60 Semantic network.

Source: Illustration courtesy of Institute for Photogrammetry and GeoInformation, University of Hannover.

Examples for such rules are:

if Landsat band 7 > Landsat band 5, then vegetation;
if radar tone is dark, then smooth surface.

It is also possible to incorporate texture parameters into the classification process. Characteristic texture parameters for an image region are:

- the autocorrelation function.
- Fourier transforms.
- grey level occurrence.

An example of a multispectral classification of a multispectral image (shown in Plate 3) is shown in Plate 4.

Multitemporal and multisensors data fusion

The prerequisite to multitemporal and to multisensoral data fusion is the geometric registration of the images to be fused. Simple examples are overlays of classified low-resolution data with high-resolution panchromatic image data. Such overlays are also effective with digitized map data.

A comparison of multitemporal data is generally done for change detection. Here again a knowledge-based approach is advantageous.

Remote sensing applications

A great number of satellite systems have provided satellite imagery for remote sensing applications in different disciplines. Table 2.11 gives a summary of the most commonly used remote satellite systems with optical sensors.

In addition to these satellites, a great number of images have been acquired through shorter duration missions by Russian photographic camera systems from Kosmos satellites and the MIR space station, e.g. the Russian KFA 1000, KVR 1000, TK 350 cameras and the German MOMS-2P digital sensor flown on the space shuttle and on MIR.

An example image is shown in Plate 5, colour section.

Table 2.12 lists the past and present radar satellites.

These images are available from nationally and internationally operating space agencies, such as NASA, NOAA, ESA, CNES, NASDA, ISRO or their vending agencies (e.g. USGS, Spot Image, Space Imaging, Eurimage, DigitalGlobe, etc.). The data cost still differs greatly at present due to an unstable market.

Meteorological data, though reduced in quality, are available over the Internet free of charge. Original resolution meteorological images can be

Table 2.11 Commonly used optical remote sensing satellites

Type	Name	Orbit	Repeat cycle	Swath	h	Resolution	Bands	Country	Application
Meteoro-logical	Meteosat	geostationary	30 min	half spheric	36000 km	5 km		ESA	meteorology, climate
	GOES-1	75° W						USA	
	GOES-2	135° W						USA	
	GSM	140° E						Japan	
	Insat							India	
	Meteor 3							Russia	
	FY 2	105° E						China	
Meteoro-logical	NOAA	polar	12 h	2394 km	705 km	1 km	5 0.58–21.5 µm	USA	climate
	DMSP	polar		2400–3000 km		3 km	6	USA	military
	5 satellites	polar					6 0.31–0.38 µm	Russia	climate (TOMS, ozone)
	Meteor P								
Earth resources	Landsat (1–3) MSS	sun-synchr. 9:30 at 40°	18 d	185 km	918 km	80 m	3–4	USA	1972–1984
	Landsat (4–5) TM	10:30 at 40°	16 d	185 km	705 km	30 m (thermal 120 m)	7	USA	1982, 1984
	Landsat 7 TM	10.00 at 40°	16 d	185 km	705 km	15 m pan 30 m MS	7	USA	1999
Earth resources	Spot P 1–4	sun-synchr. 10:00 at 40°	26 d	60 km	832 km	10 m pan 20 m MS	1 3	France	1986 1993
Earth resources	JERS 1 OPS			75 km	568 km	20 m	7	Japan	1992
Earth resources	IRS 1 A, B	sun-synchr. 9:25 at equator	22 d to 24 d	141 km	904 km 774 km	MS 36.6 m pan 5.6 m MS 23.5 m WIFS 188 m	3 1 3 2	India	1988, 1991 1995, 1997
	IRS 1 C, D								
Cartographic	Ikonos 2	sun-synchr.		11 km	677 km	1 m pan	1	USA	1999
	EROS A1	sun-synchr.		12 km	480 km	4 m MS	1	Israel	2000
	Quickbird	sun-synchr,		8 km	450 km	1.8 m pan 0.6 m pan 2.4 m MS	1	USA	2001

Table 2.12 Historical and present radar satellites

Name	Year	Inclination of orbit	Swath	h	Resolution	Polarization	Country
Seasat	1978	72°	100 km	790 km	40 m	HH 23.5 cm	USA
SIR-A	1981	50°	50 km	250 km	38 m	HH 23.5 cm	USA
SIR-B	1984	58°	40 km	225 km	25 m	HH 23.5 cm	USA
SIR-C	1994	51°	30 to 60 km	225 km	variable 13–26 m	multiple HH, HV, VH, VV, 23.5 cm, 58 cm, 3.1 cm	USA
ERS 1/2	1991, 1995	polar	100 km	785 km	30 m	VV 5.7 cm	ESA
JERS 1	1992	polar	75 km	568 km	18 m	HH 23.5 cm	Japan
Almaz	1991	polar	50 to 100 km	350 km	15 m variable	HH	Russia
Radarsat	since 1995	polar	50 to 500 km	800 km	up to 10 m	HH 5.7 cm	Canada
Envisat (ASAR)	2002	polar	100 km	800 km	12.5 m	HH, VV	ESA

obtained at reproduction cost. Medium- and high-resolution images have a weak to strong commercial component, depending on the policies of the space agency maintaining the satellite system. Privately funded commercial systems charge full price.

Global applications, therefore, use low-cost, low-resolution imagery, which is easily obtainable. Regional and local applications requiring higher resolutions rely on substantial imagery purchases. There, remote sensing competes with other data acquisition methods with respect to obtainable quality, cost and evaluation effort and time.

Project-based research applications were easiest to be realized. The present focus of applications is to concentrate on organized data acquisition and analysis programs depending on the socio-economic priorities to be placed on applications made possible by public or industrial funding.

We will now consider the situation in the major application areas of remote sensing,

Meteorology and climatology

Atmospheric sciences study the different layers of the earth's atmosphere:

- the troposphere, from 0 km to 20 km altitude.
- the stratosphere, from 20 km to 50 km altitude.
- the mesosphere, from 50 km to 80 km altitude.
- the thermosphere, from 80 km to 300 km altitude.

The 'weather zone' is the troposphere, which is of direct meteorological interest. However, the other zones also affect weather and climate.

Ozone

The earth's ozone shield extends from an altitude of 25 km to 60 km. It absorbs or reflects most of the ultraviolet light, so that only minimal amounts of ultraviolet reach the earth's surface.

NASA has launched an ultraviolet sensor, TOMS (Total Ozone Mapping Spectrometer), which observes the ozone layer in six bands between 312.5 mm and 380 mm wavelength. From these six bands, three pairs can be formed. They determine transmission minus absorption of UV energy. This ratio is a measure of ozone concentration. TOMS was carried on the US satellite Nimbus 7 from 1978 to 1993, on the Russian satellite Meteor 3 from 1991 to 1994 and on a special satellite since 1996. TOMS detected a rapidly deteriorating ozone concentration over the south polar regions in 1992 and 1993, which created great public interest. The cause could have been the eruption of the volcano Mt Pinatubo in 1991, even though aerosols produced by human activity may also have played a part. Since then, ozone measurement has been a major global remote sensing application.

The European Space Agency (ESA) launched an ozone sensor on the ERS-2 satellite in 1995. Plate 6 in the colour section shows the global ozone concentrations on a particular day (left) and a monthly average (right).

Cloud mapping

The first US imaging satellite launched in April 1961 was Tiros 1, which made it possible to observe clouds. Today geostationary satellites permit a daily weather watch following the movement of cloud patterns. The satellites Meteosat (over Africa), GOES 1 (over Venezuela), GOES 2 (over Hawaii), GMS (over the Philippines) and Insat (over the Indian Ocean) gather images every 30 minutes in the visible and thermal range. These may be geocoded and used in animations commonly shown on television programmes.

The data collected at an interval of 6 hours may be used to determine parameters of the radiative transfer model, into which the distribution of clouds, water, ice and snow and the land mass is entered.

The combination of visual and thermal bands permits the visual separation of clouds, water, ice, snow and land. If combined with atmospheric non-remote sensing measurements, a radiative transfer model can be arrived at. Plate 7 in the colour section shows the thermal GOES-1 image of a hurricane.

Rainfall

The measurement of rainfall is of great meteorological interest. The source of worldwide rainfall data are rain gauges, which are very scarcely distributed over the globe. If no rain gauge data are available over a region, data from Meteosat, GOES, etc. relating to thermal bands, can be used to determine cloud temperature. Cold clouds with $<235°K$ temperature give an indication of possible rainfall.

On the US military DMSP satellite, four wavelengths from 0.35 to 1.55 cm are provided for passive microwave sensing in two polarizations (HH, VV). The images of 55 km ground resolution permit the derivation of a scattering index, indicative of rainfall.

Wind

Since wind drives ocean currents, scatterometers can measure the roughness of the sea to estimate wind vectors. Radar images are also able to detect roughness parameters.

Weather prediction

Terrestrial measurements for weather forecasting can easily be combined with remote sensing data for cloud motion, the estimated precipitation and the measurement of surface temperature.

Other phenomena detected from images are:

- the analysis of snow cover.
- the location and the motion of tropical storms.
- the detection of fog and its dissipation.

Climate studies

Climate studies become possible by the comparison of NOAA-AVHRR aggregates on a seasonal and annual basis.

Oceanography

The geodetic aspects to be studied are ocean heights, as determined in the ESA-ERS1 Topex mission by radar altimeters. When related to the geoid, the data permit the derivation of ocean height (see Plate 8 in the colour section).

Let us now consider some phenomena of direct interest to remote sensing.

Ocean productivity

The main objective is to detect organic substances, such as phytoplankton, which are important for fisheries. It contains chlorophyll, which can be differentiated from suspended sediments prevalent near the coast and transported by estuaries (see Figure 2.61 and Plate 9 in the colour section).

Sediments reflect mainly in red. Therefore a blue/green ratio can indicate the chlorophyll concentration at sea. The actual concentration can be calibrated by in-situ measurements.

The satellites Nimbus 7 and Orbview 2 carried the sensor SeaWIFS for observation in eight channels at 1 km resolution with a swath of 2800 km. Of particular interest is the observation of sea surface temperature, which is made available at weekly intervals (see Plate 10).

Ocean currents

Ocean currents are visible along the coast because of plumes of suspended matter. In the mid-ocean the radiant temperature, which can be measured day and night by NOAA-AVHRR, shows the distribution of ocean currents. Global thermal phenomena, like El Niño, can be monitored by NOAA satellites (see Plate 11).

Radar images are also an indicator that can help to determine the level of surface roughness, since currents produce small waves.

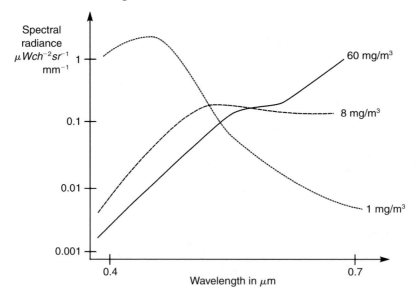

Figure 2.61 Spectral chlorophyll response at sea.

Sea ice

The principal objective of the Canadian Radarsat satellite is to be able to follow sea ice motion in the Arctic areas. In multitemporal mode the images can be observed in stereo, giving an indication of the direction of flow (see Figure 2.62).

The ESA-ERS1/2 satellites have been able to classify Arctic sea ice. The thermal band six of Landsat TM can distinguish ice temperatures. Thin ice is warmer than thick ice.

The surface roughness of sea ice can be measured by non-imaging radar scatterometers, flown from aircraft at altitudes below 1 km.

Bathymetry

Bathymetry at sea is generally made by sounding from ships, which have, however, difficulties in navigating in shallow areas. In these areas, remote sensing from aircraft and satellites can help to assess water depth, even though, depending on the turbidity of the water, light penetration and reflection is generally limited to not more than 10 m water depth. Figure 2.63 illustrates the transmission in water, which is best for the green band. A blue/green ratio can therefore be used for an assessment of water depth in shallow areas. Plate 12 in the colour section shows the composition of the Wadden Sea near Wilhalmeshaven, Germany, at low tide.

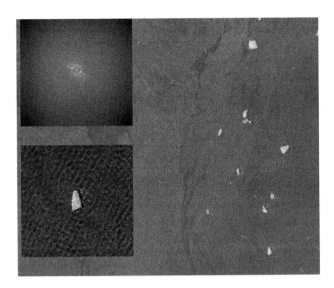

Figure 2.62 Radarsat image of icebergs on the west coast of Antarctica with wave patterns around the icebergs.

Source: ERS-1, © ESA, processed by DLR, courtesy of DLR, Oberpfaffenhofen.

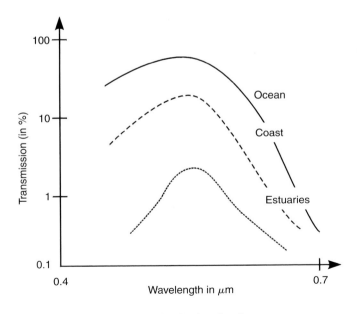

Figure 2.63 Water penetration in the visual spectrum.

Environment

Remote sensing concentrates on the detection of environmental pollution.

Hazardous waste

The task to identify hazardous waste is to find its location, to map it and to monitor it. If identification is not directly possible by visual inspection, the health of vegetation over the waste area can be a good indicator for hidden waste. A ratio of infrared/red is a good discriminator. Waste areas will usually have a low infrared/red ratio.

Plumes

Thermal pollution in water can be monitored by thermal infrared scanners operated from aircraft or by multispectral optical sensors from satellites. Plate 13 in the colour section shows the thermal pollution in the River Po, and Plate 14 shows the sediment concentration caused by the River Po.

Oil spills

Of particular environmental interest are oil spills at sea and in coastal regions. Due to its fluorescent properties, oil floating on water has a high reflectance in the ultraviolet region (0.3 μm to 0.4 μm). An aircraft scanner operated from low altitude can be used to determine the thickness of the oil film on water.

On a more regional and global level, radar images permit the location of oil slicks. They dampen the natural roughness of ocean water. They are thus observable due to their low backscatter in the images. On a local level, laser scanners are able to differentiate between oil types. Thermal images may also be of help, since the oil surface is cooler than the water.

Non-renewable resources

Locating non-renewable resources is the task of geological and geophysical exploration.

Mineral exploration

For the prospecting of minerals, the methodologies of structural geology are very important. Geological prospecting is difficult in areas covered by vegetation, however, it becomes easier in the dry belts of the globe.

Ore deposits are usually associated with structure zones. The interpretation of lineaments in the rocks permits the visible determination of fractures. Digital directional filtering of the images helps to better recognize these fractures. Fractures are often also combined with hydrothermal activity. Therefore, thermal aircraft scanners are useful in this respect. To identify the mineral composition of outcrops, hyper-spectral scanners, such as AVIRIS, offer special possibilities to distinguish minerals of outcrops.

Oil exploration

Remote sensing plays an initial role in oil exploration. The analysis of Landsat and radar images to determine the extent of sedimentary basins is usually the first step in oil exploration, even though a whole slate of geophysical prospecting methods must be applied to focus in on potential exploration locations. These are:

- aero magnetic surveys.
- gravity surveys.
- reflection seismology.
- drilling.

Renewable resources

The task of making an inventory and then monitoring renewable resources is intimately connected with the establishment of geographic information systems. It consists of compiling base data by standard mapping procedures, of supplementing these data by geocoded remote sensing data and of adding attribute data from a multitude of sources.

Remote sensing has the advantage of quick acquisition of up-to-date images at adapted spatial and spectral resolution without limitations of costly ground access. It has the disadvantage that not all desired categories of information can be extracted from the images.

Land cover – land use

Land cover describes the physical appearance of the earth's surface, while land use is a land right related category of economically using the land. Remote sensing concentrates on observing land cover.

Land cover consists of classifiable terrain objects for which different governmental base data providers have implemented a number of object-oriented classification schemes. The UN Food and Agricultural Organization has devised a hierarchical land cover catalogue suitable for the application of remote sensing and applicable to developing continents, such as Africa. It has been implemented in the Africover project separating between ninety different land cover classes in different East African countries.

The German Surveys and Mapping administration have devised an object-oriented land cover catalogue with the major categories:

1000   text information
2000   settlements
3000   transportation
4000   vegetation
5000   hydrology and water
6000   topographic elevation data
7000   administrative boundaries

These major categories are subdivided into sub-classes. Many of the sub-classes can be identified and monitored with 85 per cent accuracy by remote sensing at the global and regional or even at the local level. Figure 2.64 shows the result of a land cover classification from an airborne multi-

Figure 2.64 Land cover classification from a Daedalus Scanner image.
Source: Daedalus, © DLR, courtesy of DLR, Oberpfaffenhofen.

spectral scanner. As shown in Plate 15 in the colour section, the result is significantly enhanced if it is merged with vector information from a GIS.

Vegetation

Of particular significance is the monitoring of vegetation. On a global level, this is done from NOAA-AVHRR images obtainable twice a day. Many of these images contain clouds, which have to be eliminated from the images. This is possible by taking all of the images of a 10-day or 14-day period and substituting cloudless pixels into the data set.

While the original 1 km resolution provides too much data, a reduced resolution of 4 km or 16 km can provide a global vegetation index data set. This is possible by the use of the ratio for the normalized difference vegetation index, NDVI:

$$NDVI = \frac{Infrared - Red}{Infrared + Red}$$

The calculated NDVI for each pixel can attain a value between 1 and −1.

Green vegetation has an NDVI of about +0.7 (shown in red in Plate 16 in the colour section), while water, barren lands and clouds have an NDVI of about −0.3, shown in blue.

After radiometric calibration, the consideration of atmospheric scattering and geocoding are applied, the NDVI gives a clear indication of seasonal vegetation changes. On the northern hemisphere greenness tends to rise in May, with a peak in July and a decrease until September. Yearly comparisons for the respective months can be useful for comparing crop estimates from year to year.

Another use of the NDVI is in the monitoring of tropical vegetation and the depletion of forests, or of agricultural crops (see Plate 17 in the colour section).

Natural hazards

Earthquakes

Seismic risks are not directly observable by remote sensing. However, active faults may be visible in Landsat images, and the plate movements along these faults can be monitored by radar interferograms.

Landslides

Fresh landslides are observable in radar images.

Land subsidence

Land subsidence can be interpreted from images by a change in drainage patterns and an observation of vegetation anomalies.

Volcanoes

Volcanic eruptions are associated with clouds of ash, slope changes and mudflows. Changes in topography can be monitored by radar interferometry, but it is also possible to use thermal images from Landsat TM and from aerial scanners to study changes in heat emission.

Floods

Floods can easily be monitored with radar images. Figures 2.65 and 2.66 show multitemporal images of a river flood.

Figure 2.65 Radarsat flood image of the River Oder.

Source: © Radarsat 1997, GAF 1997, courtesy of GAF Remote Sensing and Information Systems, Munich.

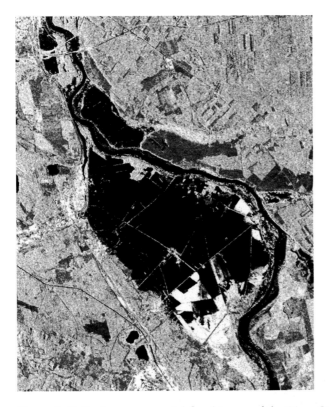

Figure 2.66 Radarsat sequential flood image of the River Oder.

Source: © Radarsat 1997, GAF 1997, courtesy of GAF Remote Sensing and Information Systems, Munich.

Forest and grass fires

In fire monitoring, three stages are important:

- the determination of fire hazards: an index with data indicating humidity, wind speed, cloud cover, ground temperature and green composition of the land can be formed to judge fire potential.
- after a fire has broken out NOAA-AVHRR can monitor the extent of the fire in the thermal band. A band 1, 2 and 4 combination can distinguish smoke and differentiate between burnt and unburnt areas.
- after a fire, damage assessment can be made with the help of the images. Plate 18 in the colour section shows the NOAA image of a forest fire along the Siberian–Chinese border.

3 Photogrammetry

Photogrammetry concerns itself with the geometric measurement of objects in analogue or digital images.

Evolution of photogrammetry

The use of photogrammetry is based on the possibility of optically projecting the terrain onto a flat surface, which thus recovers the image by means of a photographic emulsion or by digital sensors. After the invention of photography by Nièpce and Daguerre in 1839, the French military topographer Aimé Laussedat constructed a first camera in 1851, which permitted making measurements on photographs. In 1858 the German architect A. Meydenbauer introduced measurements on photos for the documentation of public buildings.

Single image photogrammetry

Because of the lack of efficient computing facilities, the reconstruction of objects was graphic, following the laws of the perspective, which had been developed during the Italian Renaissance in the fifteenth century.

With the exception of photographs from balloons, taken for military interpretation purposes in the battle of Solferino in 1859 and during the US Civil War, the standard application was terrestrial, due to the lack of a suitable aerial photographic platform.

The simplest form of restitution of single images was by rectification. Rectification could be applied to plane surfaces.

The general projective relations between the coordinates in two planes in their arbitrary coordinate systems $x'y'$ (image plane) and x, y (object plane) are:

$$x_i = \frac{a_1 x'_i + a_2 y'_i + a_3}{a_7 x'_i + a_8 y'_i + 1}$$

$$y_i = \frac{a_4 x'_i + a_5 y'_i + a_6}{a_7 x'_i + a_8 y'_i + 1}$$

This means that the coordinates of four arbitrary points in image and object define the coefficients a_1 to a_8. Based on this graphical or numerical rectification, procedures can be developed for plane surfaces (flat terrain, house walls). Figure 3.1 shows the residual differences, $\Delta r'_E$, due to topography after rectification.

The graphical restitution was extensively used in the nineteenth century. In the 1920s, special optical rectifiers were built. These had to satisfy two optical conditions: first, the lens equation:

$$\frac{1}{a} + \frac{1}{b} = \frac{1}{f}$$

with a being the image distance and b the projection distance of the image rectified and f the focal length of the optics of the rectifier.

Second, the so-called Scheimpflug condition, which stated that sharpness could only be reached if the plane of the image, the plane of the objective and the plane of the projection would intersect along a straight line in space.

Special rectification devices fulfilling these conditions automatically, using mechanical devices, were, however, not built until the mid-1920s.

These also included additional advantages: if the two arbitrary coordinate systems in the image plane and in the projection plane could be related to each other in such a way that their origins were determined by

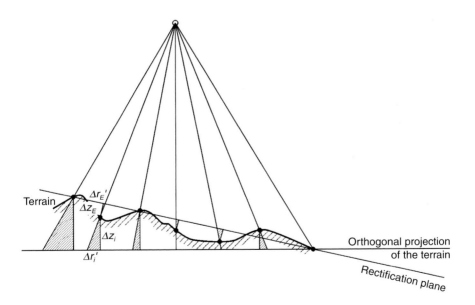

Figure 3.1 Errors in rectification.

the projection centre, then the general projective equations between the two planes could be simplified to:

$$x_i = \frac{a_1 x'_i + a_2 y'_i + 1}{a_5 x'_i + a_6 y'_i + 1}$$

$$y_i = \frac{a_3 x'_i + a_4 y'_i + 1}{a_5 x'_i + a_6 y'_i + 1}$$

Then the rules of perspective geometry can be applied. This permits the execution of a rectification with the coordinates of only three points known in image and object. The mechanical rectifiers required the centring of the images by fiducial marks. Furthermore, a shift of the image plane, d, depending on the focal length of the objective of the rectifier and the focal length of the camera taking the photograph was required.

$$d = \left(\frac{f^2_{\text{image}}}{f^2_{\text{rectifier}}} - 1 + \frac{a^2}{b^2} \right) \cdot \frac{f_{\text{rectifier}}}{2} \cdot tg\nu$$

with ν being the angle between objective and projection plane.

The procedure of three-dimensional restitution consisted of measuring image coordinates. When relating these to the projection centre identified by fiducial marks of the camera, horizontal and vertical angles to identifiable objects could be derived graphically or numerically. With the terrestrial coordinates of the exposure station determined by ground surveys, and with the exposure directions measured and set in photo-theodolites, the coordinates of the objects points could be found. Sebastian Finsterwalder used this method of photo-topography for the survey of a glacier in 1889. Photogrammetry in this simple form was an added tool to ground survey procedures in inaccessible areas (see Figure 3.2).

Analogue stereo photogrammetry

It was the stereoscopic measurement principle, developed around 1900, which permitted the automation of the restitution process. The stereocomparator of Carl Pulfrich in Germany (1901) and of Fourcade in South Africa (1901) permitted the deduction of spatial information through the measurement of observed image parallaxes (see Figure 3.3).

In order to assist stereo viewing and stereo measurement, the terrestrial photographic exposures had to be forced into a rigid configuration, preferring the normal case (see Figure 3.4).

In 1907, Eduard van Orel developed a mechanical plotting device for the reconstruction of rays measured by stereoscopic principles, the Zeiss–Orel Stereoautograph used since 1909 in the survey of Alpine

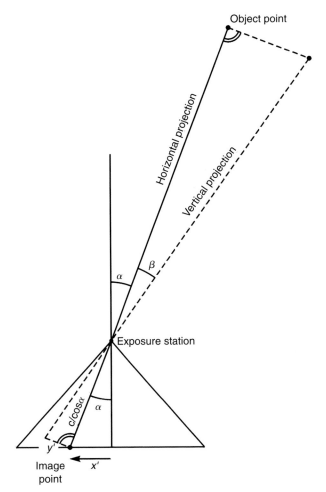

Figure 3.2 Horizontal and vertical angles derived from photocoordinates.

regions. But the more general use of photogrammetry was introduced because of the availability of controllable aerial platforms. The motorized aeroplane of the Wright brothers (from 1903) soon became an accepted photogrammetric platform, which was used extensively for interpretive purposes during World War I.

In 1915, Otto Messter developed the aerial survey camera, which permitted a systematic survey of the terrain by near vertical aerial photographs. Even though Sebastian Finsterwalder had, already in 1903, found a mathematical restitution procedure of two images taken from balloons, the lack of computing aids prevented use of an analytical reconstruction of the images.

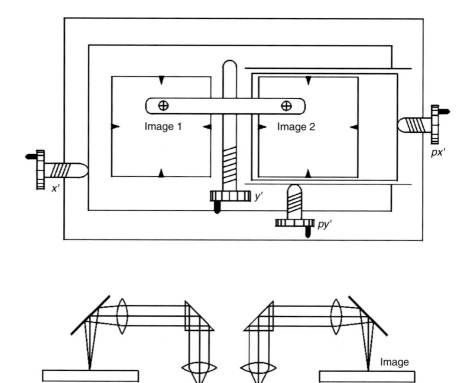

Figure 3.3 Pulfrich stereo comparator.

Also in 1915, Max Gasser patented and constructed an optical stereo-projection device which could orient two images relative to each other and which could orient both with respect to the terrain. The modified instrument was introduced into the market by the Carl Zeiss Company in 1933, under the name of 'Multiplex'. After the end of World War I, the optical industry in Europe began to construct optical and mechanical devices for the stereo reconstruction of overlapping images. Initiators of this development were R. Hugershoff in Dresden (1919), Bauersfeld of Zeiss in Jena (1921), E. Santoni and U. Nistri in Italy (1921), G. Poivilliers in France (1923), and H. Wild in Switzerland (1926), with E.H. Thompson in England and Bausch and Lomb in the USA to follow.

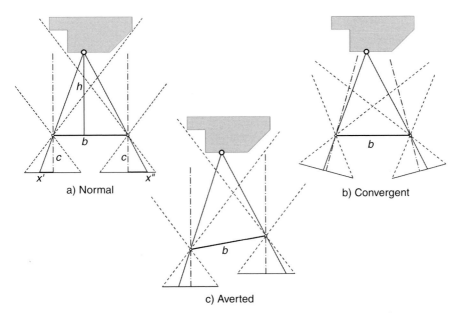

Figure 3.4 Normal, averted and convergent survey configurations of terrestrial photogrammetry.

The reconstruction of the geometry of aerial photographs was by optical means or by mechanical means, or a combination of both. The stereo measurement was made possible by an identical measuring mark inserted in the optical path or by a light mark on a projection table (see Figure 3.5).

Typical examples of this instrumental development are shown in Figures 3.6 and 3.7.

All analogue stereo restitution instruments are attempting to measure object coordinates x, y, z corresponding to a geometry of vertical photographs (see Figure 3.8).

In vertical photos, the following relations between image coordinates $x'y'$ of the first photo and the image coordinates x'', y'' of the second photo separated by the air base, b, are valid:

$$x = -\frac{b}{f}\, x'$$

$$y = -\frac{b}{f}\, y'$$

Identical floating mark for stereo observation

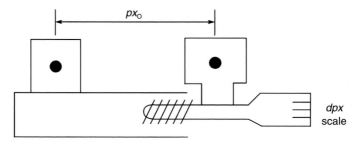

For anaglyphs, alternating shutters and for polarization filters

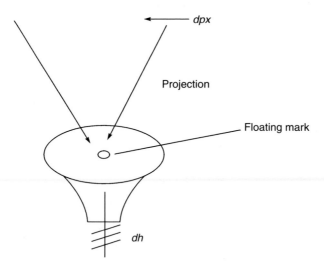

Figure 3.5 Stereo measurement.

and

$$h = -\frac{bf}{px} = -\frac{b \cdot f}{(x' - x'')}$$

The horizontal parallax, px, is therefore a measure of height differences.

For vertical photos, the vertical parallax $py = (y' - y'')$ is equal to zero in normal case images. Both image points lie in an epipolar plane.

Figure 3.9 shows in the upper part the geometric displacements of a vertical photograph due to height differences, Δh. The lower part shows horizontal displacements due to inclination of the photo by the angle, ν.

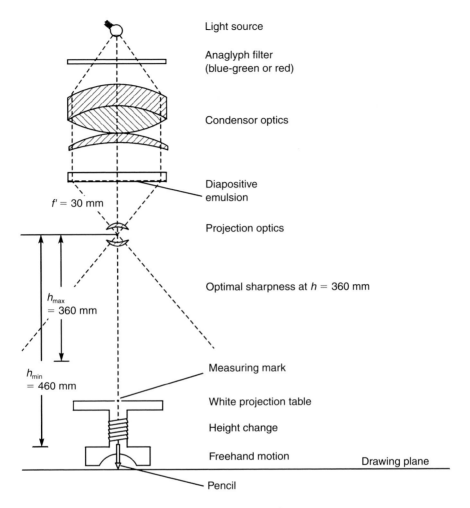

Light source

Anaglyph filter
(blue-green or red)

Condensor optics

Diapositive
emulsion

Projection optics

Optimal sharpness at $h = 360$ mm

Measuring mark

White projection table

Height change

Freehand motion

Drawing plane

Pencil

$f' = 30$ mm

h_{max}
$= 360$ mm

h_{min}
$= 460$ mm

Figure 3.6 Optical stereo restitution instrument multiplex.

Stereo restitution devices permit the changing and the measuring of coordinates x, y and z (or h) in the 'stereo model'. But they must also restore the tilt of the photographic camera and the shifts of the exposure station at the time of the exposure. To do this the plotting instruments allow a rotation of their optical or mechanical cameras around the flight axis by the angle, ω, around the vertical axis by the angle, κ, and perpendicular to it by the angle, ϕ. dx_o is a translation along the base, b, dz_o a translation along the vertical and dy_o perpendicular to it.

If a symmetrical grid of nine points in a vertical photograph is changed

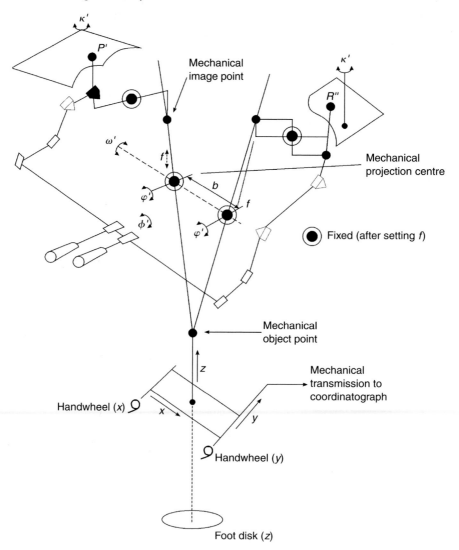

Figure 3.7 Mechanical stereo restitution instrument, Wild A8.

in orientation by these six elements, dx_o, dy_o, dz_o, $d\varphi$, $d\omega$ or $d\kappa$, then the nine points will change in position as indicated in Figure 3.10.

These effects can be used in the orientation procedures of aerial photographic diapositives in the stereo plotting instrument. In a stereo model consisting of two aerial photos overlapping by about 60 per cent, six major locations relevant for an orientation can be defined as shown in Figure 3.11.

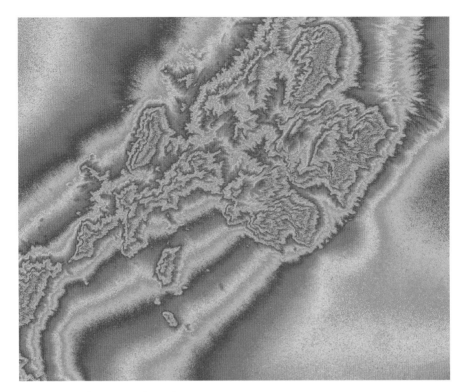

Plate 1 Interferogram of a radar image.

Source: SRTM/X-SAR © DLR, courtesy of DLR, Oberpfaffenhofen.

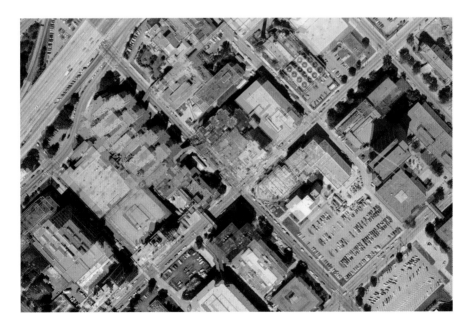

Plate 2 Anaglyph of two overlapping aerial photographs.

Source: Stereo imagery subset of Louisville, KY; data provider: Photo Science, Inc.; illustration provider: ERDAS Inc., the Geographic Imaging unit within Leica Geosystems GIS & Mapping Division, Atlanta, GA.

Plate 3 Multispectral image.

Source: Landsat 7 false colour composite: © USGS 2000, GAF 2000, courtesy of GAF Remote Sensing and Information Systems, Munich.

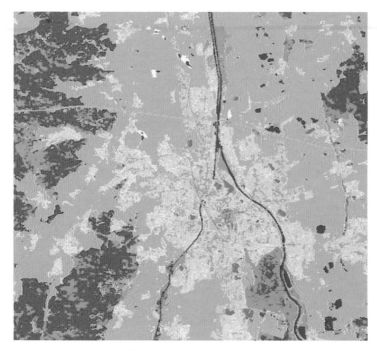

Plate 4 Multispectral classification.

Source: Land cover classification: © GAF 2000, courtesy of GAF Remote Sensing and Information Systems, Munich.

Plate 5 IRS 1 C/D image of Munich Airport in Germany with 5.6 m
panchromatic resolution fused with 23.5 m multispectral resolution.

Source: IRS 1D-PAN/LISS image: Munich © SI/Antrix/Euromap 1999, GAF 2000, courtesy
of GAF Remote Sensing and Information Systems, Munich.

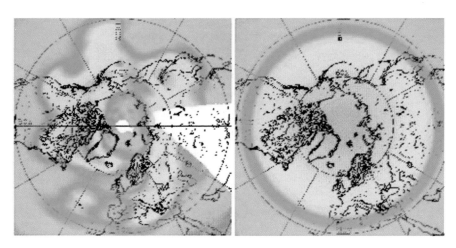

Plate 6 Global ozone concentration during the global ozone monitoring
experiment (GOME) (blue = 200 Dobson units; red = 500 Dobson units).

Source: ERS-2/GOME, © DLR, courtesy of DLR, Oberpfaffenhofen.

Plate 7 Hurricane thermal image from GOES-1.

Source: Meteosat-3 MVISSR, © Eumetsat, processed by DLR, courtesy of DLR, Oberpfaffenhofen.

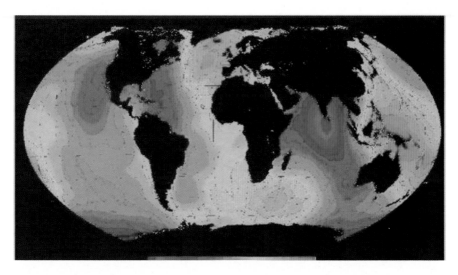

Plate 8 Radar altimetry from the Topex Mission by ERS-1 (violet = 10.5 m; red = +83 m for differences between WGS 84 ellipsoid and geoid plus sea surface topography).

Source: © DLR, courtesy of DLR, Oberpfaffenhofen.

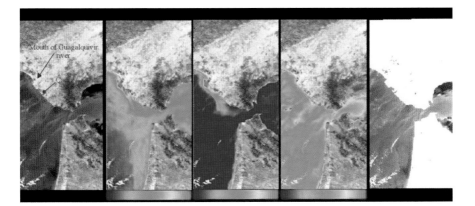

Plate 9 Separation of pigments, sediments and aerosols on MOS Images from IRS
for the Strait of Gibraltar (left: multispectral image followed by the
separation of pigments (chlorophyll concentrations), sediments and
aerosol concentrations; and right: a black and white image for the
separation of clouds over the sea; pigments in µg (l: blue = 0, green = 3,
red = 6; sediment: blue = 0, red = 5; aerosol optical thickness: blue = 0,
red = 1)).

Source: MOS-IRS, © DLR, courtesy of DLR, Oberpfaffenhofen.

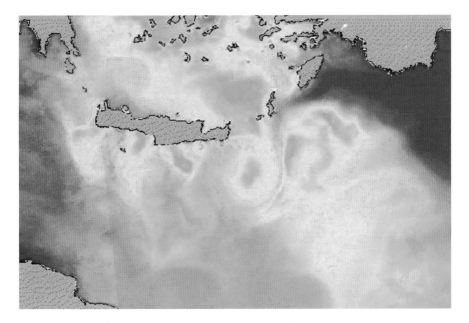

Plate 10 Sea surface temperature in the Eastern Mediterranean (weekly mean
temperature, summer 1994: dark red = 30°C, blue = 23°C).

Source: NOAA-AVHRR, © DLR, courtesy of DLR, Oberpfaffenhofen.

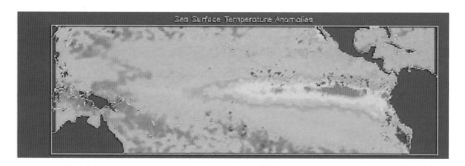

Plate 11 El Niño (sea surface temperature anomaly by ERS along track scanning radiometer: monthly mean, October 1997).

Source: Image from CEOS CDROM 98, courtesy of CNES, Toulouse.

Plate 12 The composition of the Wadden Sea, near Wilhelmshaven in Germany, at low tide.

Source: IRS 1D-LISS Image, SI/Antrix/Euromap 1998, GAF 2000, courtesy of GAF Remote Sensing and Information Systems, Munich.

Plate 13 Thermal pollution in the River Po.
Source: Daedalus, DLR, courtesy of DLR, Oberpfaffenhofen.

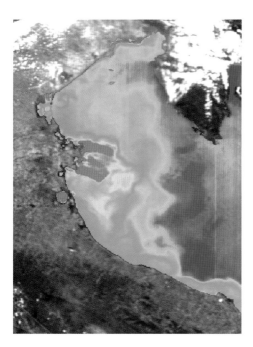

Plate 14 Sediment concentration at sea caused by the Po in the Adriatic Sea.
Source: MOS-IR, © DLR, courtesy of DLR, Oberpfaffenhofen.

Plate 15 Land cover merged with GIS vector data.
Source: Daedalus., © DLR, courtesy of DLR, Oberpfaffenhofen.

Plate 16 NDVI of the Oder Region from MOS on IRS computed from channels 7 (651 nm) and 11 (869 nm).
Source: MOS-IRS, © DLR, courtesy of DLR, Oberpfaffenhofen.

Plate 17 NDVI from a Daedalus Scanner for an agricultural area.

Source: NDVI image, © SI/Antrix/Euromap 1997, GAF 1999, courtesy of GAF Remote Sensing and Information Systems, Munich.

Plate 18 Forest fire at the Siberian–Chinese boundary.

Source: Image from CEOS CDROM 98, courtesy of CNES, Toulouse.

Plate 19 The German topographic map 1:25 000 derived from ATKIS.

Source: Map courtesy of LGN, Hannover.

Plate 20 Superposition of old cadastre and digital orthophoto in Croatia.

Source: Map courtesy of State Geodetic Administration (DGU), Republic of Croatia, Zagreb.

Plate 21 Urban utility network.

Source: Map courtesy of Ministry of Municipal Affairs and Agriculture, Doha, Qatar, C.E.
Dept., Drainage GIS Section, and is used herein with permission by ESRI, Redlands and the
Ministry of Municipal Affairs and Agriculture, Doha, Qatar.

Plate 22 Extraction of buildings from a GIS.

Source: Image courtesy of Phoenics GmbH, Hannover.

Plate 23 Addition of building façade texture, orthophoto texture for the ground and derivation of rooftypes from DSM height differences for buildings.

Source: Image courtesy of Phoenics GmbH, Hannover.

Plate 24 Governmental, residential and commercial holdings in the city of Bangkok.

Source: Map courtesy of Metropolitan Electricity Authority, Thailand, and is used herein with permission by ESRI, Redlands, CA, and Metropolitan Electricity Authority for educational purposes.

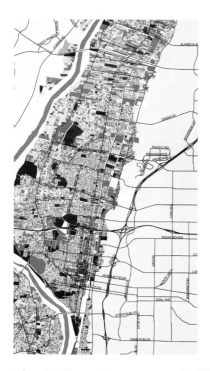

Plate 25 Ownership type map of a US city.

Source: Map courtesy of Middle Rio Grande Conservancy District, Albuquerque, New Mexico, and is used herein with permission by ESRI, Redlands, CA, and the Middle Rio Grande Conservancy District.

Plate 26 Beetle infestation in New York City.

Source: Map courtesy of City of New York Parks and Recreation, New York, NY, and is used herein with permission by ESRI, Redlands, CA, and City of New York.

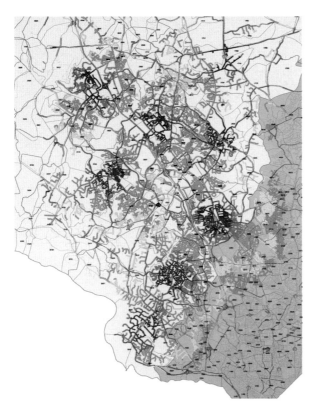

Plate 27 Distances to the nearest fire station.

Source: Map courtesy of Montgomery County, Rockville, Maryland, and is used herein with permission by ESRI, Redlands, CA, and Montgomery County.

Plate 28 Frequency of burglaries by city blocks in a US city.

Source: Graphic image supplied courtesy of ESRI-Arc View GIS, and is used herein with permission by ESRI, Redlands, CA.

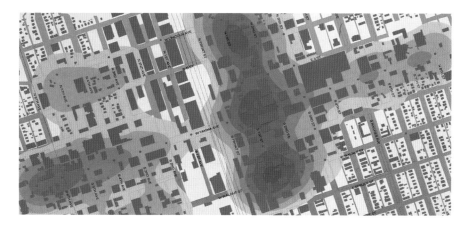

Plate 29 Industrial air pollution in a city.

Source: Map courtesy of City of Yakima, Yakima, Washington, and is used herein with permission by ESRI, Redlands, CA and the City of Yakima.

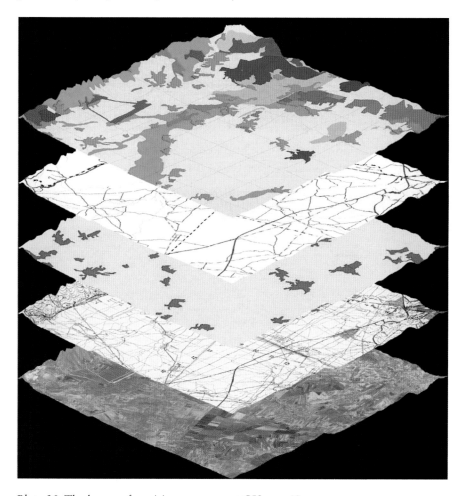

Plate 30 The layers of a crisis management GIS over Kosovo.

Source: Landsat, source: P. Reinartz *et al.*, in PGM 2000/3, processed by DLR, courtesy DLR, Oberpfaffenhofen.

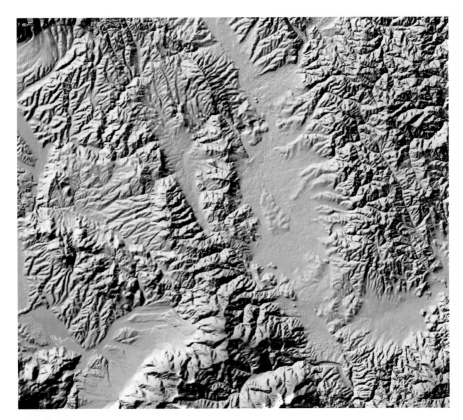

Plate 31 The DEM over Kosovo, generated from radar interferometry.

Source: © DLR and JRC, processed by DLR, courtesy of DLR, Oberpfaffenhofen.

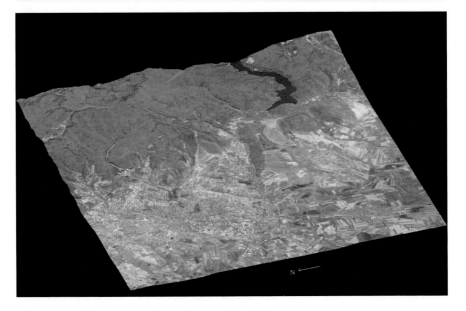

Plate 32 Satellite imagery draped over the DEM.

Source: © DLR and JRC, processed by DLR, courtesy of DLR, Oberpfaffenhofen.

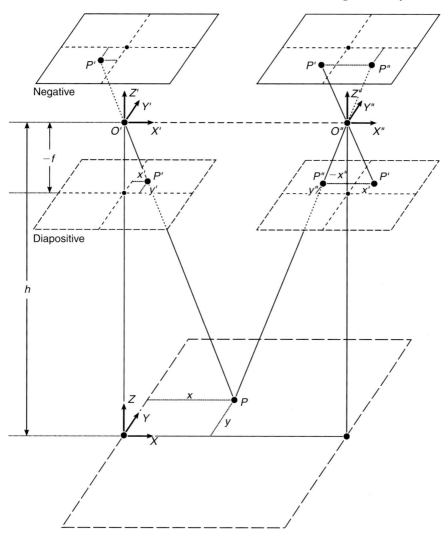

Figure 3.8 Geometry of vertical photographs.

Stereo restitution in the plotting instrument can then be carried out in the following steps.

Interior orientation consisting of centring the aerial photographic dia-positives in the plate holders by help of the fiducial marks and by setting the principal distance of the instrument to the focal length of the camera. This assures a perspective solution can be applied.

Relative orientation of one photo with respect to the other is needed to permit stereo viewing in epipolar planes. Due to the fact that aerial photos,

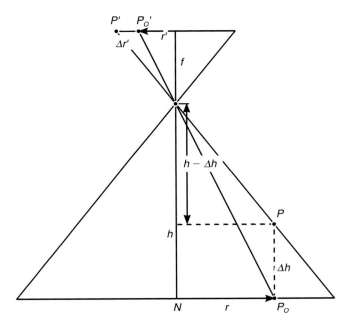

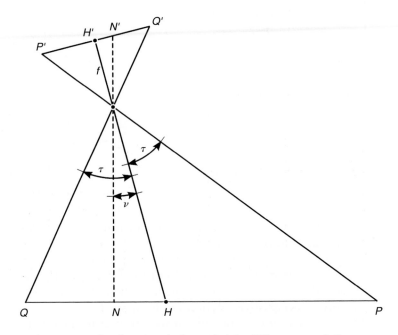

Figure 3.9 Displacements of a photograph due to height differences and tilt.

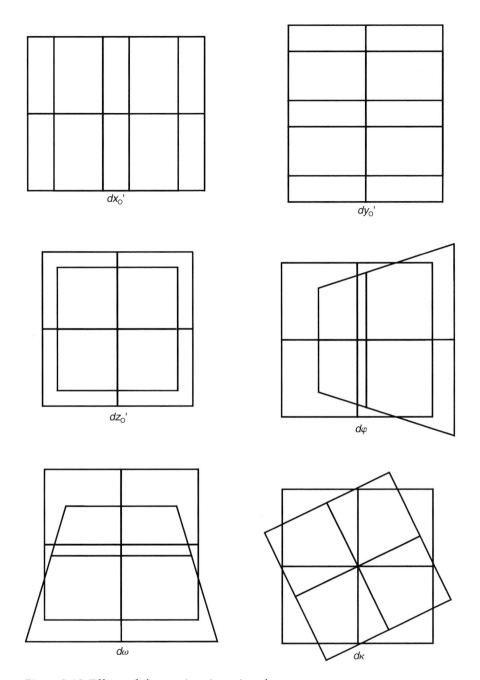

Figure 3.10 Effects of changes in orientation elements.

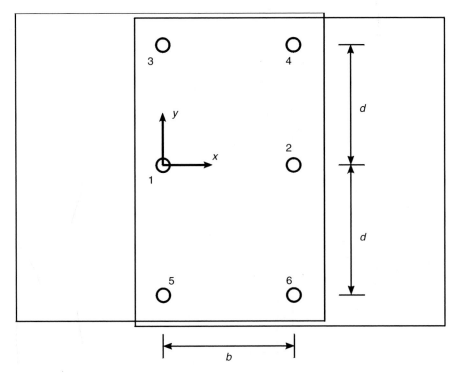

Figure 3.11 Von Gruber points of relative orientation in a stereo model.

as a rule, are only *nearly* vertical, corresponding points projected from the two photos will not intersect in one point as shown in Figure 3.12, causing horizontal and vertical parallaxes of the corresponding rays. While the horizontal parallax can be removed by height, the vertical parallax is caused by a faulty relative orientation.

The effect of faulty relative orientation on the heights in a stereo model is shown in Figure 3.13.

As early as 1930, Otto von Gruber had introduced an iterative optical–mechanical relative orientation procedure. At the six von Gruber points shown in Figure 3.11, the vertical parallaxes in y are eliminated by the elements shown, while the horizontal parallaxes in x only represent elevation differences between these points. It is also possible to calculate orientation changes for the cameras numerically, if the plotting instrument has dials to change them. The vertical parallaxes in y are measured at the von Gruber points and the orientation changes are then computed and introduced (see Figure 3.14).

In independent pair relative orientation, the rotations of both projectors

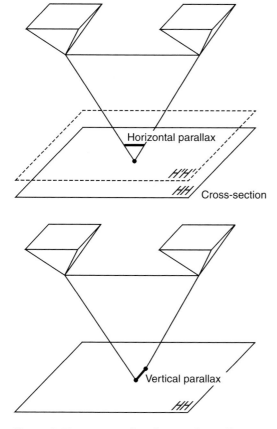

Figure 3.12 Horizontal and vertical parallax.

are used to obtain a *y*-parallax free stereo model at the von Gruber points. In dependent pair relative orientation, rotations and translations of only one projector are used. The functions $f(py_i)$ for dependent pair analogue numerical orientation are, for example:

$$bz'' = -\frac{b}{2d}(py_6 - py_4)$$

$$\varphi'' = -\frac{b}{2bd}[(py_5 - py_5) - (py_6 - py_4)]$$

$$\varphi'' = -\frac{b}{4d^2}(2py_1 + 2py_2) - py_3 - py_4 - py_5 - py_6)$$

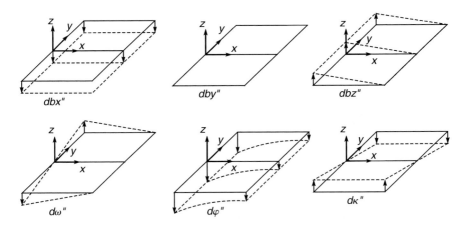

Figure 3.13 Model deformation due to faulty relative orientation.

$$\kappa'' = -\frac{1}{3b}[(py_2 - py_1) + (py_4 - py_3) + (py_6 - py_5)]$$

by″ does not need to be calculated, since it can be visually removed.

Absolute orientation reorients the relatively oriented image pair in space to a ground reference given by control points. One spatial distance between two given points and three elevations spanning a triangle are required for this transformation (see Figure 3.15).

After the orientation has been accomplished, the operator of the plotting instrument can trace the visible details identifiable in the stereo model. Thus mapping of streets, rivers and creeks, settlements and houses, and vegetation boundaries is carried out on a map manuscript. Setting the floating mark of the instrument at a certain even height, contours characterizing the terrain can be drawn.

Plotting instruments even provided the opportunity for control extension along the flight strip by means of analogue aerial triangulation. As long as identical transfer points in the von Gruber locations were measured in each stereo model, the intersecting rays could be used to reduce the number of required control points in a photogrammetric block.

A great variety of optical and, particularly, mechanical stereo instruments have been produced since World War II by Swiss, German, Italian, French, Russian and British optical manufacturers. The models differed slightly in the ways in which the spatial reconstruction was accomplished. The difference was due mainly to the need not to violate patent rights. Nevertheless, the restitution methodology remained common. With these instruments, the current state of topographic mapping in the world has

Analogue optical–mechanical orientation

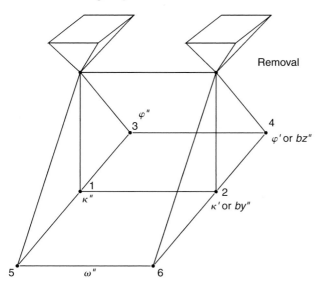

Removal

Analogue numerical orientation

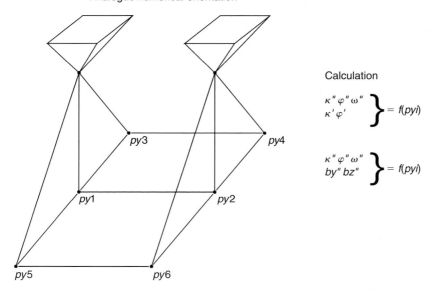

Calculation

$$\left.\begin{matrix} \kappa''\,\varphi''\,\omega'' \\ \kappa'\,\varphi' \end{matrix}\right\} = f(pyi)$$

$$\left.\begin{matrix} \kappa''\,\varphi''\,\omega'' \\ by''\,bz'' \end{matrix}\right\} = f(pyi)$$

Figure 3.14 Relative orientation.

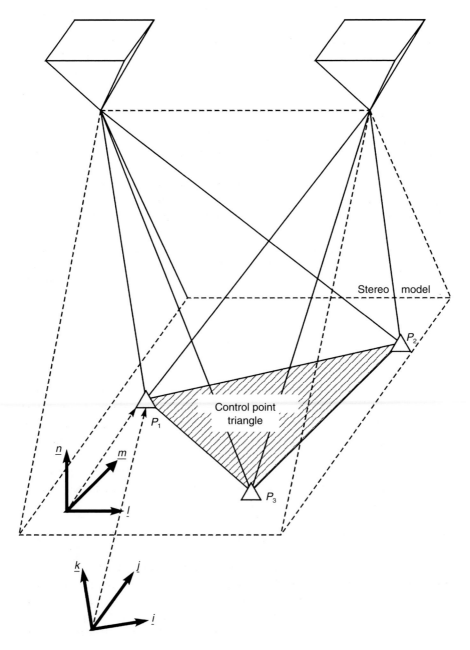

Figure 3.15 Absolute orientation.

been achieved. This was largely due to the training facilities offered by the ITC Netherlands as part of international cooperation.

In the Western hemisphere and other developed countries these analogue instruments have now been replaced by analytical and digital restitution instruments. But they continue to be used in developing countries.

Analytical photogrammetry

Starting in the 1950s, computer systems became available which permitted a rigorous treatment of the photogrammetric solution. This treatment, first applied to point measurement in images or the stereo model by a human operator, not only automated many steps of the photogrammetric restitution process, but also made photogrammetry more accurate and more reliable by use of least square adjustment and statistical testing included in the computer solutions.

Analytical photogrammetry concerns itself with the modelling of sensor geometry and its restitution. As such, it maintains its full validity even at a time when most restitution operations have turned to the use of digital images and digital image processing.

The principles of analytical and digital photogrammetry are discussed in this chapter, pages 127–182.

The advancements brought about by analytical photogrammetry can be summarized thus:

- the measurement of image coordinates of corresponding points in a photographic model or block permitted the calculation and adjustment of the bundle of rays determining the three-dimensional geometry. This solved the need for control extension in photogrammetric blocks.
- The possibility to rapidly perform digital conversion of coordinates opened the way for a new design of stereo restitution instruments in the form of analytical plotters.

The principle of the functioning of an analytical plotter is shown in Figure 3.16.

In analogue stereo plotting devices, model coordinates x, y, z were physically located by the plotter operator in the scale of the stereo model. The observation of corresponding image points was realized by an analogue projection with optical and mechanical devices responsible for the coordinate conversion into the image systems of $x'y'$ and $x''y''$ coordinates.

An analytical plotter functions in a similar way, except that the real time coordinate conversion for observation of image points is done on the basis of computation. Servomotors are used to shift the observation system to the computed positions. This has the advantage that the range of

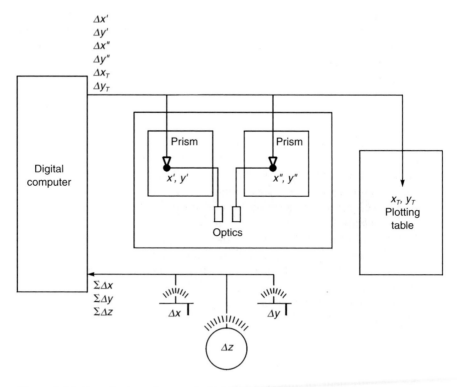

Figure 3.16 Functioning of an analytical plotter.

computations can be extended to cover geometries, which cannot be handled by the relatively simple optical and mechanical devices.

This means that the geometry of aerial photographs can be corrected on-line for lens and film distortions known from an off-line or on-line calibration process. It means further that stereo restitution can be extended into the use of unconventional imaging geometries provided by optical and microwave imaging devices.

Due to the possibility of steering the analytical plotter automatically to known *xyz* positions stored in the computer, this offers an automated facility for performing a semiautomatic interior, relative and absolute orientation, for semiautomatic or automatic point transfer from model to model and for the measurement of predetermined elevation grids. This grid determination of heights in the analytical plotter performs more rapidly than the following of contours, which is standard with analogue instruments.

All measured point coordinates are recorded in the computer files together with an appropriate code. The digital data sets acquired in analytical plotters can therefore be easily incorporated in further GIS processing operations. Analytical plotters have so far successfully competed with digital techniques for the extraction of linear features in topographic mapping, since their stereo-viewing and stereo-interpretation capabilities of the film diapositives have generally been superior to the evaluation of digital images in stereo workstations.

Both LH Systems and Z/I Imaging still market analytical plotters (see Figures 3.17 and 3.18).

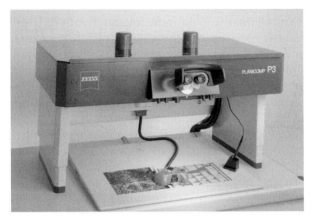

Figure 3.17 The Z/I Imaging Planicomp P3.

Source: Image courtesy of Z/I Imaging Corp., Oberkochen.

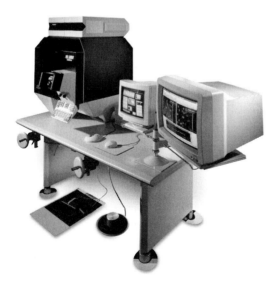

Figure 3.18 The LH Systems SD 2000/3000.

Source: Image courtesy of LH Systems (Leica Geosystems), San Diego, CA. © Leica Geosystems, 2002.

The automation capabilities of analytical plotters were also successfully incorporated into the process of differential rectification, in which the image geometry, distorted due to perspective displacements of different terrain heights, was converted into ortho-projected orthophotos.

Even though orthophoto instrumentation began its development in the early 1930s, with O. Lacmann in Germany and R. Ferber in France, and even though the US Geological Survey instituted on orthophoto mapping program in the 1950s with specially constructed analogue devices, orthophotography only became a generally accepted tool after commercial development by Zeiss in Germany, Wild in Switzerland and OMI in Italy manufactured orthophoto devices on the analytical plotter principle. Figure 3.19 shows the functioning of an analytical orthophoto device.

Analytical orthophoto printers allowed the projection of an image slit onto a drum containing photographic film. While the system permitted the exposure along the drum by moving an optical system along it, the next slits could be exposed by a stepwise shift on the drum.

The analytical plotter controlled the optical projection of the slit, changing its location on the photo, $x'y'$, its magnification, dm, and its rotation, Θ', under computer control during the scan. Analytical ortho-printers have now been replaced by digital orthorectification.

Digital photogrammetry

Digital photogrammetry makes use of digital or digitized images. This

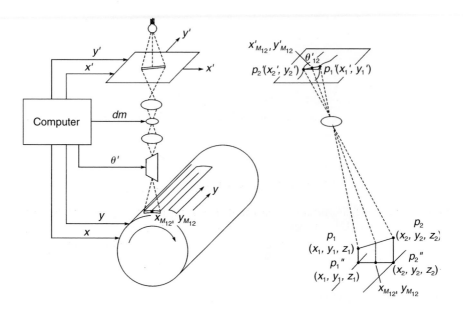

Figure 3.19 Analytical orthophoto device.

permits vastly extended automation possibilities. Even though digital photogrammetry was already experimentally realized by John Sharp of IBM in 1965, it required the development of fast computers with adequate storage facilities before it could become practical. This time arrived around 1988 when the first digital photogrammetric workstations were shown at the ISPRS Kyoto Congress.

In addition to the capabilities offered by analytical plotters, digital workstations now permit the use of newly developed automated technology such as:

- automated or automatic aerial triangulation.
- the derivation of digital elevation models by image matching techniques.
- added display and analysis techniques.
- integration into GIS systems.

A digital stereo workstation consists of the following parts:

- a central processing unit with sufficient performance of hundreds to thousand MHz.
- a 32-bit operating system.
- a large enough memory with at least 64 Mb RAM, preferably 256 Mb or 515 Mb RAM.
- a large enough storage system to store a sufficient number of images at least in the tens of GB.
- a graphics system to permit stereo viewing and measurement.
- a graphic user interface.

The prime manufacturers of digital workstations are:

- LH-Systems with SOCET-SET (see Figure 3.20).
- Z/I Imaging with Image Station SSK and with Image Station 2001 or Z4 (see Figure 3.21).
- Virtuozo.

Principles of analytical and digital photogrammetry

Fundamental to the modern treatment of photogrammetry are the tools of matrix algebra and of least square adjustment. With these tools, the problems of spatial networks and their coordinate system conversions can be efficiently treated.

Coordinate transformations between image and terrain

Basic to photogrammetric restitution is the conversion of two-dimensional

Figure 3.20 The LH Systems SOCET-SET.

Source: Image courtesy of LH Systems (Leica Geosystems), San Diego, CA. © Leica Geosystems, 2002.

Figure 3.21 The Z/I Imaging Image Station 2001.

Source: Image courtesy of Z/I Imaging Corp., Huntsville, AL (Image Station ®).

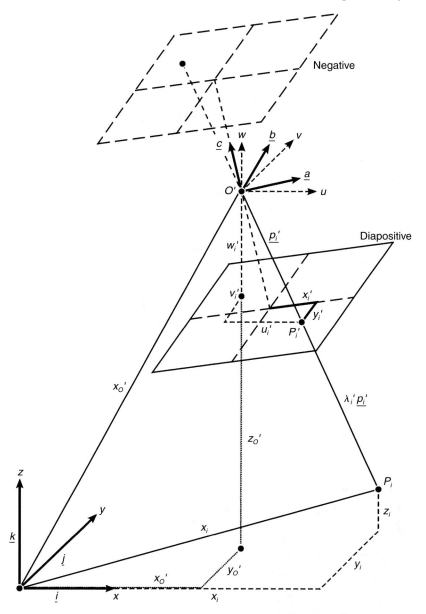

Figure 3.22 Image and object coordinate systems.

image coordinates into three-dimensional object coordinates and vice versa. The basic relations are shown in Figure 3.22.

Image coordinates and local Cartesian object coordinates

The relations can be expressed as three-dimensional vectors between the following points:

- O, the origin of the Cartesian object coordinate system, x, y, z
- P_i, the object point with its coordinates, x_i, y_i, z_i, in that system, expressed by the vector $\vec{x}_i$.
- O', the exposure station with its coordinates x'_o, y'_o, z'_o, in that system, expressed by the vector $\vec{x}'_o$.
- with these two vectors the vector OP_i forms a spatial triangle.

$O'P_i$ can be expressed by the image coordinates, x', y' and f or the vector $\vec{p}'_i$. Prerequisite for the perspective transformation is that the origin of the image coordinate system is linked to the projection centre of the exposure station, O'.

Furthermore, the image coordinate system requires a spatial rotation expressed by a three-dimensional rotation matrix, R, and a scale change, λ'_i, between image measurements and the object coordinates.

The vectors in the triangle $OO'P_i$ then can be added:

$$\vec{x}_i = \vec{x}_o + \lambda'_i R \cdot \vec{p}'_i$$

Its coordinate components are:

$$\begin{pmatrix} x_i \\ y_i \\ z_i \end{pmatrix} = \begin{pmatrix} x'_o \\ y'_o \\ z'_o \end{pmatrix} + \lambda'_i R \begin{pmatrix} x'_i \\ y'_i \\ -f \end{pmatrix}$$

For the convenience of having the axes of the object and image coordinate systems pointing in similar directions, the principal distance, f, is conventionally entered with a minus sign. Both systems are considered as orthogonal, with their unit vectors $\vec{i}, \vec{j}, \vec{k}$ and $\vec{a}, \vec{b}, \vec{c}$ perpendicular to each other, and each axis has the same scale. In this case the inverse relation does not require calculation of an inverse matrix, R^{-1} from R, but it can be expressed as its transpose, R^T.

$$\begin{pmatrix} x' \\ y' \\ -f \end{pmatrix} = \frac{1}{\lambda'_i} R^T \begin{pmatrix} x_i - x'_o \\ y_i - y'_o \\ z_i - z'_o \end{pmatrix}$$

There are various ways to define the rotational matrix, R. This can be done with the use of direction cosines of the spatial angles between the axes $x'x, x'y \ldots z'z$:

$$R = \begin{pmatrix} \cos(x'x)\cos(y'y)\cos(z'x) \\ \cos(x'y)\cos(y'y)\cos(z'y) \\ \cos(x'z)\cos(y'z)\cos(z'z) \end{pmatrix}$$

It is, however, more convenient to follow the tradition of analogue photogrammetry. In analogue instruments, the rotation around the x-axis was called ω, around the y-axis, φ, and around the z-axis, κ. The rotations R_ω, R_φ and R_κ could be formed accordingly. In plotting devices, however, the inverse relations converting object to image coordinates were preferred with the rotations R_ω^T, R_φ^T, R_κ^T.

A derivation of the coefficients of the rotations matrix R_κ^T is shown in Figure 3.23.

$$x' = x\cos\kappa + y\sin\kappa$$

$$y' = y\cos\kappa + x\sin\kappa$$

Thus the three-dimensional matrix becomes:

$$\begin{pmatrix} x' \\ y' \\ z' \end{pmatrix} = \begin{pmatrix} \cos\kappa & \sin\kappa & 0 \\ -\sin\kappa & \cos\kappa & 0 \\ 0 & 0 & 1 \end{pmatrix} \begin{pmatrix} x \\ y \\ z \end{pmatrix} = R_K^T \begin{pmatrix} x \\ y \\ z \end{pmatrix}$$

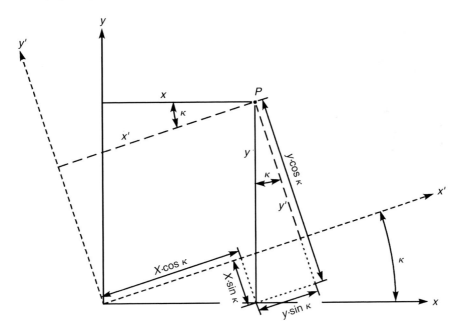

Figure 3.23 Rotational matrix derivation.

$$R_\phi^T = \begin{pmatrix} \cos\varphi & 0 & \sin\varphi \\ 0 & 1 & 0 \\ -\sin\varphi & 0 & \cos\varphi \end{pmatrix}, \text{ and}$$

$$R_\omega^T = \begin{pmatrix} 1 & 0 & 0 \\ 0 & \cos\omega & \sin\omega \\ 0 & -\sin\omega & \cos\omega \end{pmatrix}$$

For plotting instruments converting object coordinates into image coordinates, these were introduced in a sequential manner such that:

$$\begin{pmatrix} x' \\ y' \\ z' \end{pmatrix} = R_\kappa^T \cdot R_\phi^T \cdot R_\omega^T \begin{pmatrix} x \\ y \\ z \end{pmatrix}$$

The matrix multiplication of these three sequential matrices results in:

$$R^T = \begin{pmatrix} r_{11} & r_{21} & r_{31} \\ r_{12} & r_{22} & r_{32} \\ r_{13} & r_{23} & r_{33} \end{pmatrix} = \begin{pmatrix} \cos\kappa\cos\phi & \sin\kappa\cos\omega & \sin\kappa\sin\omega \\ & -\cos\kappa\sin\varphi\sin\omega & +\cos\kappa\sin\varphi\cos\omega \\ -\sin\kappa\cos\varphi & \cos\kappa\cos\omega & \cos\kappa\sin\omega \\ & +\sin\kappa\sin\varphi\sin\omega & -\sin\kappa\sin\varphi\sin\omega \\ -\sin\varphi & -\cos\varphi\sin\omega & \cos\varphi\cos\omega \end{pmatrix}$$

Since these coefficients must all represent the direction cosines of the angles between the axes, it is possible to derive the matrix, R^T, also with other sign conventions and other rotational sequences. The resultant coefficients of R^T can thus be used to calculate the respective ω, ϕ, κ values by the simple comparison of the coefficients.

The object coordinate system x, y, z is a local Cartesian coordinate system which must be related to the geodetic coordinate system based on a reference frame in use in a particular country. The relations between geodetic coordinates are available in geodetic literature.

In a single model or in a small area, the difference between the geodetic reference and the Cartesian object model is so small that it has often been neglected in practice. For a more thorough treatment, an example for converting 3° transverse mercator (Gauss–Krüger) coordinates into a local Cartesian system is given below.

Transverse Mercator and geographic coordinates

Control points are usually known in 3° Transverse Mercator coordinates, Xi, Yi. These can be transformed into geographic coordinates, φ_i, λ_i (see Figure 3.24).

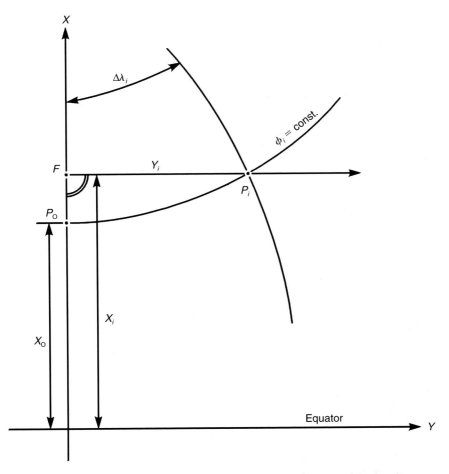

Figure 3.24 Conversion between transverse mercator and geographic coordinates.

$$\varphi_i = \varphi_F - \frac{\rho}{2N_F^2} t_F (1 + \eta_F^2) Y_i^2 + \frac{\rho}{24N_F^4} t_F (5 + 3t_F^2 + 6\eta_F^2 - 6t_F^2 - 6t_F^2\eta_F^2) Y_i^4$$

with

$$\varphi_F = V_F^2 \left[\frac{X_i}{N_F} - \frac{3}{2} \eta_F^2 t_F \frac{X_i^2}{N_F^2} + \frac{X_i^3}{N_F^3} \cdot \frac{\eta_F^2}{2} (t_F - 1 - \eta_F^2 + 5\eta_F^2 t_F)^2 \right]$$

$$V_F = \sqrt{1 + e'^2 \cos^2 \varphi_F}$$

$$N_F = \frac{a}{[(1 - e^2)\sin^2 \varphi_F]^{1/2}}$$

$$\eta_F = (e'^2 \cos^2 \varphi_F)^{1/2}$$

$$e = \sqrt{\frac{a^2 - b^2}{a^2}}, \; e' \sqrt{\frac{a^2 - b^2}{b^2}}, \; \rho^o = \frac{180^o}{\pi}$$

$$t_F = t_g \varphi_F$$

a and *b* are parameters of the chosen reference ellipsoid.
λ_F is the chosen principal meridian.
φ_F is iteratively calculated.

The inverse transformation is:

$$X_i = X_o + \frac{1}{2}\frac{N_i}{\rho^2}\cos^2 \varphi_i t_i (\lambda - \lambda_o)^2 + \frac{1}{24}\frac{N_i}{\rho^4}\cos^2 \varphi_i t_i (5 - t_i^2 + 9\eta_i^2 + 4\eta_i^4)(\lambda_i - \lambda_o)^4$$

$$Y_i = \frac{1}{\rho}N_i \cos \varphi_i (\lambda_i - \lambda_o) + \frac{1}{6}\frac{N_i}{\rho^3}\cos^3 \varphi_i (1 - t_i^2 + \eta_i^2)(\lambda_i - \lambda_o)^3$$

$$+ \frac{1}{120}\frac{N_i}{\rho^5}\cos^5 \varphi_i (5 - 18t_i^4 + 14\eta_i^2 - 58t_i^2\eta_i^2)(\lambda_i - \lambda_o)^5$$

with

$$t_i = t_g \varphi_i$$

$$N_i = \frac{a}{1 + e^2 \sin^2 \varphi_i)^{1/2}}$$

$$\eta_i = (e'^2 \sin^2 \varphi_i)^{1/2}$$

$$V_i = \sqrt{1 + e'^2 \cos^2 \varphi}$$

$$X_o = N_i \left[\frac{\varphi_i}{V_i} + \left(\frac{\varphi_i}{V_i} \right)^2 \cdot \frac{3}{2}\eta_i^2 t_i - \left(\frac{\varphi}{V_i} \right)^3 \frac{\eta_i^2}{2}(t_i^2 - 1 - \eta_i^2 - 4\eta_i^2 - t_i^2) \right]$$

Geographic coordinates and geocentric Cartesian coordinates

Geographic coordinates, φ_i, λ_i, can be converted into geocentric Cartesian coordinates (space rectangular coordinates) X'_i, Y'_i, Z'_i (see Figure 3.25).

$$X'_i = (N_i + h_i)\cos\varphi_i\cos\lambda_i$$

$$Y'_i = (N_i + h_i)\cos\varphi_i\sin\lambda_i$$

$$Z'_i = [N_i(1 - e^2) + h_i]\sin\varphi_i$$

The ellipsoidal height, h_i, of a point is composed of the orthometric height, H_i, minus the geoidal undulation, G_i:

$$h_i = H_i - G_i$$

The inverse solution is:

$$tg\lambda_i = \frac{Y'_i}{X'_i}$$

$$(N_i + h_i) = \sqrt{X'^2_i + Y'^2_i + (Z'^2_i + t_1)^2}$$

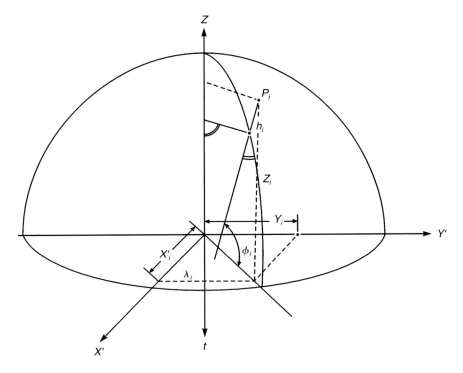

Figure 3.25 Geocentric Cartesian coordinates.

Since $t_1 = e^2 Z'^2$ and iterative solution is required using $(Z_i'^2 + t) = (N_i + h)$ $\sin \varphi_1$ with

$$\varphi_1 = \arcsin \frac{Z_i' + t_1}{(X_i'^2 + Y_i'^2 + (Z_i'^2 + t_1)^2)^{1/2}}$$

with this:

$$N_1 = \frac{a}{\sqrt{1 - e^2 \sin \varphi_1}} \quad \text{and} \quad t_2 = N_1 e^2 \sin \varphi$$

After a few iterations, φ_i is obtained as well as:

$$h_i = \sqrt{X_i'^2 + Y_i'^2 + (Z_i'^2 + t_1)^2} - N_i$$

Transformation between geocentric and local Cartesian coordinates

Transformation of the geocentric Cartesian coordinate system into a local one is possible by the following transformation:

$$\begin{pmatrix} x_i \\ y_i \\ z_i \end{pmatrix} = m \begin{pmatrix} -\sin \lambda_o & \cos \lambda & 0 \\ -\sin \varphi_o \cos \lambda_o & -\sin \varphi_o \sin \lambda_o & \cos \varphi_o \\ \cos \varphi_o \cos \lambda_o & \cos \varphi_o \sin \lambda_o & \sin \varphi_o \end{pmatrix} \begin{pmatrix} x_i' - x_o' \\ y_i' - y_o' \\ z_i' - z_o' \end{pmatrix}$$

$$= m M \begin{pmatrix} X_i' - X_o' \\ Y_i' - Y_o' \\ Z_i' - Z_o' \end{pmatrix}$$

with X_o', Y_o', Z_o', φ_o and λ_o' defining the origin of the local system. A scale factor, m, may also be introduced.

The inverse solution is:

$$\begin{pmatrix} X_i' \\ Y_i' \\ Z_i' \end{pmatrix} = \begin{pmatrix} X_o' \\ Y_o' \\ Z_o' \end{pmatrix} + \frac{1}{m} M^T \begin{pmatrix} x_i \\ y_i \\ z_i \end{pmatrix}$$

Control point coordinates may be converted by these transformations into the photogrammetric object coordinate system, in which the analytical solution is computed. Thereafter all local coordinates can be retransformed into the coordinates of the projection used.

The reason why this conversion process has been neglected in the past

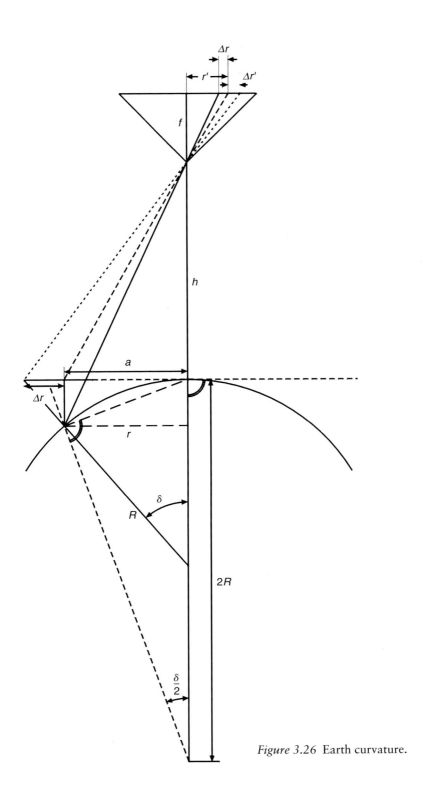

Figure 3.26 Earth curvature.

was due to the use of a great number of control points in the block between which the transformation errors have been interpolated. The same is true for the apparent height deformation due to earth curvature, d (see Figure 3.26), which amounts to about

$$d \approx \frac{r^2}{R}$$

The same holds true for atmospheric refraction, which acts in the form of a radial distortion (see Figure 3.27).

Space intersection

For the determination of spatial coordinates, at least two photographs from different exposure stations are required. It then becomes possible to apply the transformation between object coordinates x_i, y_i, z_i and photo

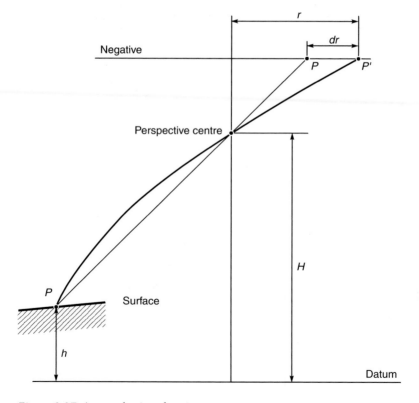

Figure 3.27 Atmospheric refraction.

coordinates of the imaged points in both photos $x_i'y_i'$ and $x_i''y_i''$ (see Figure 3.28).

$$\vec{x}_i = \vec{x}_o + \lambda_i'R'\vec{p}_i' = \vec{x}_i = \vec{x}_o'' + \lambda_i''R''\vec{p}_i'' \text{ with the components:}$$

$$\begin{pmatrix} x_i' \\ y_i' \\ z_i' \end{pmatrix} = \begin{pmatrix} x_o' \\ y_o' \\ z_o' \end{pmatrix} + \lambda_i'R' \begin{pmatrix} x_i' \\ y_i' \\ -f \end{pmatrix} = \begin{pmatrix} x_o'' \\ y_o'' \\ z_o'' \end{pmatrix} + \lambda_i''R'' \begin{pmatrix} x_i'' \\ y_i'' \\ -f \end{pmatrix}$$

If the coordinates of both exposure stations, $\vec{x}_o'$ and $\vec{x}_o''$, and their photo orientations expressed through their two rotational matrices, R' as a func-

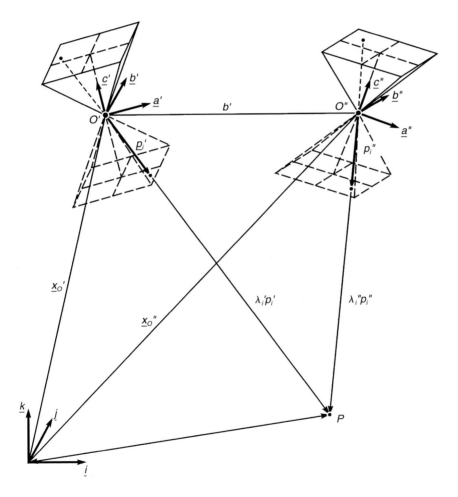

Figure 3.28 Space intersection.

tion of ω', φ', κ' and R'' as a function of ω'', φ'', κ'' are known, then two of the equations may be used to determine the unknown scale factors, λ'_i and λ''_i.

If the projections of the photo coordinate vectors, $\vec{p}'_i$ and $\vec{p}'_i$, in the directions of the object coordinate system are defined as:

$$R'\begin{pmatrix} x'_i \\ y'_i \\ -f \end{pmatrix} = \begin{pmatrix} u' \\ v' \\ w' \end{pmatrix} \text{ and}$$

$$R''\begin{pmatrix} x''_i \\ y''_i \\ -f \end{pmatrix} = \begin{pmatrix} u'' \\ v'' \\ w'' \end{pmatrix}$$

then the equations can be written as:

$$\begin{pmatrix} x_i \\ y_i \\ z_i \end{pmatrix} = \begin{pmatrix} x'_o \\ y'_o \\ z'_o \end{pmatrix} + \lambda'_i \begin{pmatrix} u'_i \\ v'_i \\ w'_i \end{pmatrix} = \begin{pmatrix} x''_o \\ y''_o \\ z''_o \end{pmatrix} + \lambda''_i \begin{pmatrix} u''_i \\ v''_i \\ w''_i \end{pmatrix}$$

The two equations with the largest coordinate differences $(x''_o - x'_o) = bx$ and $(y''_o - y'_o) = by$ may be used to solve for λ'_i and λ''_i.

$$\lambda'_i = \frac{bx \cdot v''_i - by \cdot u''_i}{u'_i v''_i - v'_i u''_i}$$

$$\lambda''_i = \frac{by \cdot u'_i - bx \cdot v'_i}{v'_i \cdot u''_i - u'_i \cdot v''_i}$$

with λ'_i and λ''_i known, the coordinates of the intersected point, $P_i(x_i, y_i, z_i)$ may be calculated.

Space resection

Unless the coordinates of the exposure station, x'_o, y'_o, and z'_o, can be directly measured by GPS and unless the components of the rotational matrix, R', as functions of the angles, ω', φ', κ', can be directly surveyed by inertial measuring units (IMUs), the orientation of a single photograph consisting of the six orientation elements, x'_o, y'_o, z'_o, ω', φ', κ', must be determined by a space resection based on known control points identifiable in the photo (see Figure 3.29).

For this, the equations of coordinate conversion between object and photo can be used:

$$\vec{x}_i = \vec{x}'_o + \lambda'_i \cdot R \cdot \vec{p}'_i$$

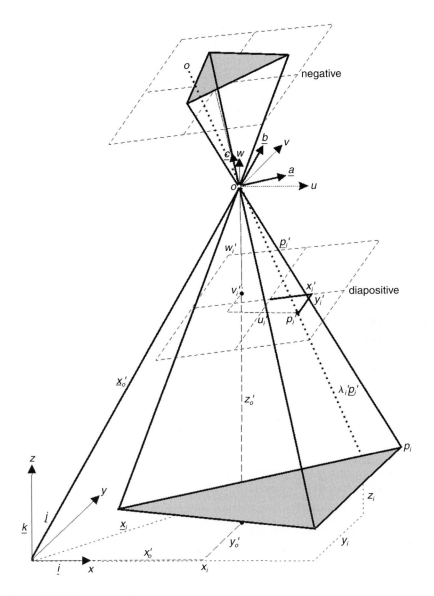

Figure 3.29 Space resection.

with its components

$$\begin{pmatrix} x_i \\ y_i \\ z_i \end{pmatrix} = \begin{pmatrix} x'_o \\ y'_o \\ z'_o \end{pmatrix} + \lambda'_i \begin{pmatrix} r_{11} & r_{12} & r_{13} \\ r_{21} & r_{22} & r_{23} \\ r_{31} & r_{32} & r_{33} \end{pmatrix} \begin{pmatrix} x'_i \\ y'_i \\ -f \end{pmatrix}$$

In these equations, the photo coordinates are the observations and the orientation elements, x'_o, y'_o, z'_o, ω', φ, κ, as well as λ'_i, the unknown.

It is therefore more convenient to use the inverse relations:

$$
\begin{pmatrix} x'_i \\ y'_i \\ -f \end{pmatrix} = \frac{1}{\lambda'_i} \begin{pmatrix} r_{11} & r_{21} & r_{31} \\ r_{12} & r_{22} & r_{32} \\ r_{13} & r_{23} & r_{33} \end{pmatrix} \begin{pmatrix} x_i - x'_o \\ y_i - y'_o \\ z_i - z'_o \end{pmatrix}
$$

Since f is a constant of the camera used, it is possible to eliminate λ'_i by dividing the first and second equation through the third. This will result in the collinearity equations, which are an expression or a condition, that image point, projection centre and object point must lie on a straight line:

$$
x'_i = -f \frac{r_{11}(x_i - x'_o) + r_{21}(y_i - y'_o) + r_{31}(z_i - z'_o)}{r_{13}(x_i - x'_o) + r_{23}(y_i - y'_o) + r_{33}(z_i - z'_o)}
$$

$$
y'_i = -f \frac{r_{12}(x_i - x'_o) + r_{22}(y_i - y'_o) + r_{33}(z_i - z'_o)}{r_{13}(x_i - x'_o) + r_{23}(y_i - y'_o) + r_{33}(z_i - z'_o)}
$$

With respect to the six unknowns, these are non-linear equations. To solve them for the required number of points, they are linearized by a Taylor series using its linear terms as a first approximation, with iterations to follow:

$$
x'_i = (x'_i)_o + \frac{\delta x'_i}{\delta x'_o} dx'_o + \frac{\delta x'_i}{\delta y'_o} dy'_o + \frac{\delta x'_i}{\delta z'_o} dz'_o + \frac{\delta x'_i}{\delta \omega'_o} d\omega' + \frac{\delta x'_i}{\delta \varphi'_o} d\varphi'_o + \frac{\delta x'_i}{\delta \kappa'_o} d\kappa'
$$

$$
y'_i = (y'_i)_o + \frac{\delta y'_i}{\delta x'_o} dx'_o + \frac{\delta y'_i}{\delta y'_o} dy'_o + \frac{\delta y'_i}{\delta z'_o} dz'_o + \frac{\delta y'_i}{\delta \omega'_o} d\omega' + \frac{\delta y'_i}{\delta \varphi'_o} d\varphi'_o + \frac{\delta y'_i}{\delta \kappa'_o} d\kappa'
$$

The differential coefficients are after differentiation of the collinearity equations as follows:

$$
\frac{\delta x'_i}{\delta x'_o} = -\frac{1}{D}(-x'_i r_{13} - f r_{11})
$$

$$
\frac{\delta x'_i}{\delta y'_o} = -\frac{1}{D}(-x'_i r_{23} - f r_{21})
$$

$$
\frac{\delta x'_i}{\delta z'_o} = -\frac{1}{D}(-x'_i r_{33} - f r_{31})
$$

$$\frac{\delta x'_i}{\delta \omega'_o} = -\frac{1}{D}\{x'_i[(y_i - y'_o)r_{33} + (z_i - z'_o)r_{23}] + f[-(y_i - y'_o)r_{31} + (z_i - z'_o)r_{21}]\}$$

$$\frac{\delta x'_i}{\delta \varphi'_o} = -\frac{1}{D}\begin{bmatrix} x'_i[-(x_i - x'_o)\cos\varphi' + (y_i - y'_o)\sin\varphi'\sin\omega' - (z_i - z'_o)\sin\varphi'\cos\omega'] \\ +f[-(x_i - x'_o)\cos\kappa'\sin\varphi' - (y_i - y'_o)\cos\kappa'\cos\omega' + (z_i - z'_o)\cos\kappa'\cos\varphi'\cos\omega'] \end{bmatrix}$$

$$\frac{\delta x'_i}{\delta \kappa'_o} = -\frac{1}{D}\{f[(x_i - x'_o)r_{12} + (y_i - y'_o)r_{22} + (z_i - z'_o)r_{32}]\}$$

$$\frac{\delta y'_i}{\delta x'_o} = -\frac{1}{D}(-y'_i r_{13} - f r_{12})$$

$$\frac{\delta y'_i}{\delta y'_o} = -\frac{1}{D}(-y'_i r_{23} - f r_{22})$$

$$\frac{\delta y'_i}{\delta z'_o} = -\frac{1}{D}(-y'_i r_{33} - f r_{32})$$

$$\frac{\delta y'_i}{\delta \omega'_o} = -\frac{1}{D}\{y'_i[-(y_i - y'_o)r_{33} + (z_i - z'_o)r_{23}] + f[-(y_i - y'_o)r_{32} + (z_i - z'_o)r_{22}]\}$$

$$\frac{\delta y'_i}{\delta \phi'_o} = -\frac{1}{D}\begin{bmatrix} y'_i[-(x_i - x'_o)\cos\varphi' + (y_i - y'_o)\sin\varphi'\sin\omega' - (z_i - z'_o)\sin\varphi'\cos\omega'] \\ +f[-(x_i - x'_o)\sin\kappa'\sin\varphi' - (y_i - y'_o)\sin\kappa'\cos\omega'\sin\omega' - (z_i - z'_o)\sin\varphi'\cos\varphi'\cos\omega'] \end{bmatrix}$$

$$\frac{\delta y'_i}{\delta \kappa'_o} = -\frac{1}{D}\{f[-(x_i - x'_o)r_{11} - (y_i - y'_o)r_{21} - (z_i - z'_o)r_{31}]\}$$

with the abbreviation

$$D = r_{11}(x_i - x'_o) + r_{23}(y_i - y'_o) + r_{33} \times (z_i - z'_o)$$

The first approximations for $(x'_i)_o$ and $(y'_i)_o$ are derived from a flight plan and the assumptions of vertical photography:

$$(x'_i)_o = -f\frac{x_i - (x'_o)_o}{z_i - (z'_o)_o}$$

$$(y'_i)_o = -f\frac{y_i - (y'_o)_o}{z_i - (z'_o)_o}$$

In matrix notation, the equation system can be written as:

$$\underset{i,j}{A}\,\vec{x}_{j,1} = \vec{\ell}_{i,1}$$

in which A_{ij} is the coefficient matrix, $\vec{x}_{j,1}$ the six orientation unknowns and $\vec{\ell}_{i,1}$ the observations i. In order to determine the six orientation parameters, a minimum of three control points are required. They will permit the setting up of six observation equations for these three points:

$$x'_i, y'_i$$

These permit solving for the unknowns:

$$x_{6,1} = A^{-1}_{6,6}\ell_{6,1}$$

The inverse of the coefficient matrix can be calculated if $|A| \neq 0$. This is the case if the three control points form a spatial triangle. If more than three control points are available, then the number of observations will be greater than six (e.g. eight observations for four points), and the determination of the orientation parameters will be subject to a least squares adjustment.

For the equations of all point residuals, v'_{x_i}, v'_{y_i} must be introduced, so that:

$$v_{i,1} = A_{ij}, x_{j,1} - \ell_{i,1} \text{ or in the example:}$$

$$\underset{8,1}{v} = \underset{8,6}{A} \ \underset{6,1}{x} - \underset{8,1}{\ell}$$

A least squares solution is obtained if the residuals are made to a minimum. This is the case if:

$$\frac{\delta v_i^T v_i}{\delta x_j} = 0 = A_{j,i}^T A_{i,j} x_{j,i} - A_{j,i}^T \ell_{i,1}$$

Thus:

$$x_{j,1} = (A_{j,i}^T A_{i,j})^{-1} A_{j,i}^T \ell_{i,1} \text{ or}$$

$$x_{6,1} = (A_{6,8}^T A_{8,6})^{-1} A_{6,8}^T \ell_{8,1}$$

The approximations for x_j can now be replaced by these values and a new iteration can start. The procedure is repeated until no changes in x_j become noticeable.

With the orientation parameters calculated in this manner for all photographs, a space intersection for the determination of object coordinates can follow without the need for a relative and absolute orientation of the stereo model.

Aerial triangulation

The problem of space resection can be extended to include the simultaneous orientation of multiple exposure stations, which are common in a photogrammetric block. For this, it is necessary to include observations of transfer points, which have been identified and measured in adjacent photos. The object coordinates of these points must be added to the equation system as three additional unknowns per transfer point so that the linearized collinearity equations become:

$$x'_i = (x'_i)_o + \frac{\delta x'_i}{\delta x'_o}dx'_o + \frac{\delta x'_i}{\delta y'_o}dy'_o + \frac{\delta x'_i}{\delta z'_o}dz'_o + \frac{\delta x'_i}{\delta \omega'}d\omega' + \frac{\delta x'_i}{\delta \varphi'}d\varphi'_o + \frac{\delta x'_i}{\delta \kappa'}d\kappa'$$

$$+ \frac{x'_i}{\delta x_i}dx'_o + \frac{\delta x'_i}{\delta y_i}dy'_o + \frac{\delta x'_i}{\delta z_i}dz'_o$$

$$y'_i = (y'_i)_o + \frac{\delta y'_i}{\delta x'_o}dx'_o + \frac{\delta y'_i}{\delta y'_o}dy'_o + \frac{\delta y'_i}{\delta z'_o}dz'_o + \frac{\delta y'_i}{\delta \omega'}d\omega' + \frac{\delta y'_i}{\delta \varphi'}d\varphi'_o + \frac{\delta y'_i}{\delta \kappa'}d\kappa'$$

$$+ \frac{\delta y'_i}{\delta x_i}dx_i + \frac{\delta y'_i}{\delta y_i}dy_i + \frac{\delta y'_i}{\delta z_i}dz_i$$

These new differential coefficients become, after differentiation of the collinearity equations:

$$\frac{\delta x'_i}{\delta x_i} = -\frac{1}{D}(x'_i \cdot r_{13} + fr_{11})$$

$$\frac{\delta x'_i}{\delta y_i} = -\frac{1}{D}(x'_i \cdot r_{23} + fr_{21})$$

$$\frac{\delta x'_i}{\delta z_i} = -\frac{1}{D}(x'_i \cdot r_{33} + fr_{32})$$

$$\frac{\delta y'_i}{\delta x_i} = -\frac{1}{D}(y'_i \cdot r_{13} + fr_{12})$$

$$\frac{\delta y'_i}{\delta y_i} = -\frac{1}{D}(y'_i \cdot r_{23} - fr_{22})$$

$$\frac{\delta y'_i}{\delta z_i} = -\frac{1}{D}(y'_i \cdot r_{33} + fr_{32})$$

These equations can be applied to all measured points of the block in preparation for a bundle block adjustment. The configuration of photographic bundles and their relationship to stereo models is shown in Figure 3.30.

Figure 3.31 illustrates the application to a block consisting of four photos each in three flight strips (total of twelve photos) with twenty-eight transfer points in the von Gruber locations.

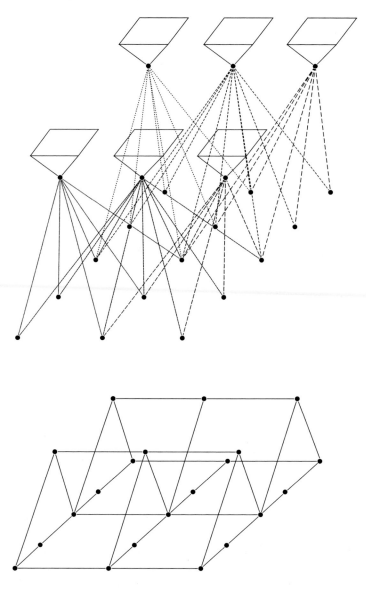

Figure 3.30 Photos and models of a block.

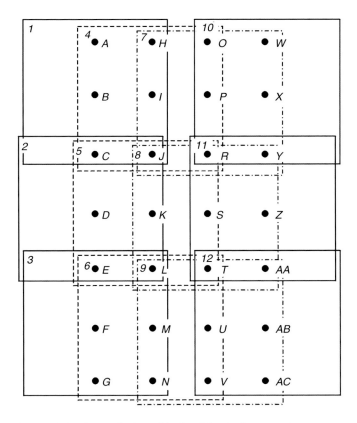

Figure 3.31 Example for a block of 4 × 3 photographs.

There are $12 \times 6 = 72$ orientation unknowns and there are $28 \times 3 = 84$ transfer point unknowns in the block (total 156 unknowns, x_j).

Photos 1, 2, 3, 10, 11 and 12 contain the measurements $x'_i y'_i$ of six transfer points (twelve observation equations each) and photos 4, 5, 6, 7, 8 and 9 contain nine transfer points (eighteen observation equations each). This results in 180 observation equations for this relatively small block:

$$v_{180,1} = A_{180,156} x_{156,1} - \ell_{180,1}$$

requiring to invert a matrix of 156×156 equations:

$$x_{156,1} = (A^T_{156,180} \cdot A^T_{180,156})^{-1} (A^T_{156,180} \cdot \ell_{180,1})$$

The matrix is invertible, if $|A^T A|^{-1} \neq 0$, requiring the knowledge of a minimum number of ground control points. A typical control point

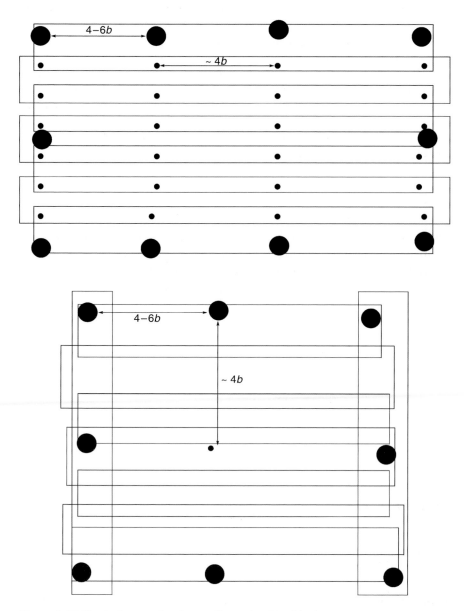

Figure 3.32 Typical control point configuration in a block.

configuration in a block is shown in Figure 3.32. The big dots represent horizontal control points and the small dots represent vertical control points.

If some of the transfer points used are ground control points, then their coordinate unknowns can be directly eliminated. To reach the condition, four well-placed ground control points are needed. Thus, twelve coordinate unknowns can be eliminated to reach a solution.

It is, however, preferable to leave the equation system untouched and to add the ground control values as additional observations, with the advantage that they can be appropriately weighted with regard to the measurements in the photos. For the four ground control points, there will be twelve additional equations.

Each observation equation can be assigned a weight, p_i. It is related to the variance of the observation:

$$p_i = \frac{1}{\sigma_i^2}$$

Photo coordinate measurements can be assigned a weight commensurate with the standard deviation, σ_i of measurement (e.g. 8 μm). The ground coordinates of control points can be entered with a weight:

$$p_x = \frac{1}{\sigma_x^2}$$

commensurate with their accuracy of point determination.

The least squares adjustment principle will then be:

$$v^T p v = (Ax - \ell)^T p (Ax - \ell) = \min$$

with the solution:

$$x = (A^T p A)^{-1} A^T p \ell$$

yielding a

$$\sigma_o^2 = \frac{v^T p v}{n - u}$$

with n being the number of observations and u the number of unknowns.

The use of weighted added observations is also possible for direct measurement of exposure stations via GPS and of camera orientations via IMU.

Modern bundle block adjustment programs, such as BLUH established at the University of Hannover, are capable of treating 6000 photo orientations and 200 000 ground points in a simultaneous adjustment.

As the computational power needed for matrix inversion increases with the third power of the size of the matrix, the symmetry of the arrangement of observation helps in matrix partitioning with the possibility to invert partial inverses more efficiently to permit a solution for the 600 000-plus unknowns.

The matrix *A* for the example selected in Figure 3.33 is shown in Figure 3.33.

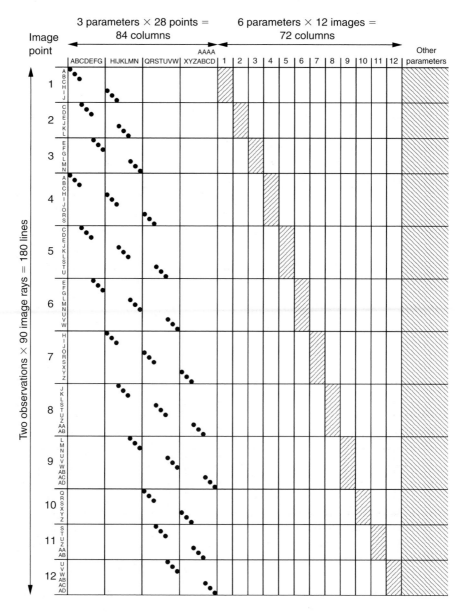

Figure 3.33 Error equation matrix.

The error equations do not show correlations for the orientation elements of different exposure stations; the point coordinates are also uncorrelated. Only a limited number of parameters are correlated. When the normal equations, $A^T pA$, are formed, the correlations between points and photos become visible (see Figure 3.34).

This permits the grouping of matrix A for the error equations into three parts:

$$v'_i = (A_1 A_2 A_3) \begin{pmatrix} x_1 \\ y_2 \\ z_3 \end{pmatrix} - \ell_i$$

A_1 contains the matrix part for the point unknowns, A_2 the matrix part for

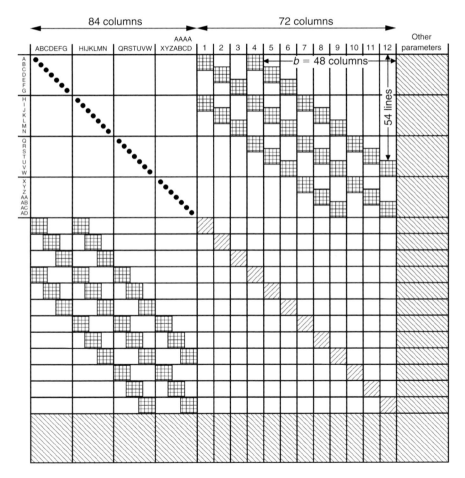

Figure 3.34 Normal equation matrix $A^T pA$.

the orientation unknowns, and A_3 any additional parameters. The normal equation matrix, $A^T p A$, thus is composed of the following groups:

$$
\begin{pmatrix}
A_1^T A_1 & A_1^T A_2 & A_1^T A_3 \\
A_2^T A_1 & A_2^T A_2 & A_3^T A_3 \\
A_3^T A_1 & A_3^T A_2 & A_3^T A_3
\end{pmatrix}
\begin{pmatrix}
x_1 \\
x_2 \\
x_3
\end{pmatrix}
=
\begin{pmatrix}
A_1^T \ell \\
A_2^T \ell \\
A_3^T \ell
\end{pmatrix}
$$

Parts of this matrix do not need to be fully inverted in the solution. $A_2^T A_2$ only consists of 3×3 elements for unknown points which can easily be inverted. This results in additions to $A_2^T A_2$, which has uncorrelated 6×6 elements. The additions of the partially inverted $A_1^T A_1$ matrix and the matrix parts $A_2^T A_1$ or $A_1^T A_2$ structure matrix $A_2^T A_2$ into a bandwidth structure around the diagonal. The inversion is possible by recursive partitioning. Finally, only $A_3^T A_3$ needs to be fully inverted.

The additional parameters x_3 consist of self-calibration parameters, which can be modelled to determine systematic deformations common for each image. A typical set of additional parameters can include radial lens distortion, expressible by a polynomial, tangential lens distortion and film deformations (affinity, skewness).

The error equations can, for example, be augmented by the following terms:

$$
v'_{x_i} = \ldots a_1 x'_i (r'_i - r'_o) + a_2 x'_i (r'^3_i - r'^3_o) + a_3 x'_i (r'^5_i - r'^5_o) + a_4 x'_i (r'^7_i - r'^7_o)
$$
$$
+ a_5 x'_i \cos 2\alpha_i + a_6 x'_i \sin 2\alpha_i - a_7 y'_i \cos 2\alpha_i - a_8 y'_i \sin 2\alpha_i - a_9 x'_i \ldots
$$

$$
v'_{y_i} = \ldots a_1 y'_i (r'_i - r'_o) + a_2 y'_i (r'^3_i - r'^3_o) + a_3 y'_i (r'^5_i - r'^5_o) + a_4 y'_i (r'^7_i - r'^7_o) +
$$
$$
a_5 y'_i \cos 2\alpha_i + a_6 y'_i \sin 2\alpha_i + a_7 x'_i \cos 2\alpha_i + a_8 x'_i \sin 2\alpha_i - a_9 y'_i + a_{10} x'_i \ldots
$$

α_i determines the axis of symmetry for asymmetric and tangential lens distortion and $r'_i = \sqrt{x'^2_i + y'^2_i}$ the radial distance; r'_o is a chosen fixed value, e.g. 100 mm to which radial distortion is referred.

A solution is first attempted without the use of additional parameters obtaining a particular σ_o for the solution. Thereafter, the parameters will be included. This should reduce σ_o. It is advantageous to statistically test the additional parameters to judge whether they contribute to the improvement.

For GPS flights, when observations for the exposure stations are added, further additional parameters modelling the effect of cycle slips between strips may be included. The bundle block adjustment is a rigid solution, which can be applied to any photogrammetric orientation and point determination problem. However, some simpler approaches are applicable for the orientation of single stereo models, such as relative and absolute orientation.

Relative orientation

Relative orientation of a photograph with respect to the first photo is made if the exterior orientation of the first photo to ground coordinates (e.g. by space resection) has not yet been accomplished. Therefore, one makes the choice of local object coordinate system, the model coordinate system, which is based on the assumption that the first image is a vertical image with $x'_o y'_o z'_o \omega' \varphi' \kappa'$ all zero. The calculation of the orientation parameters of the second photo, $x''_o y''_o z''_o \omega'' \varphi'' \kappa''$, needs to be determined, so that stereoscopic viewing along epipolar lines is possible. The determination of x''_o is not yet of importance, since the model coordinate system has an arbitrary scale.

Figure 3.28 shows that the vectors b', $\lambda'_i p'_i$ and $\lambda''_i p''_i$ form a triangle O', O'' and P. Addition of these vectors $b' = \lambda'_i p'_i - \lambda''_i p''_i$, requires the knowledge of model coordinates formed by a space intersection. A solution is possible without these, using the vectors b', $\bar{p}'_i$ and $\bar{p}''_i$ only without the scale factors. The relations in the triangle may be expressed by the vector product of the three vectors, stating that the tetrahedron spanned between these vectors must be zero.

$$(b' \times p'_i) \cdot p''_i = 0$$

This is known as the coplanarity equation. As these vector operations may only be performed in one chosen coordinate system, e.g. that of b', the components of p'_i and $\bar{p}''_i$ must be expressed in their projections to the axes of the base vector. Camera 1 has the rotational matrix, R' and camera 2, R''.

$$\bar{p}'_i = \begin{pmatrix} u'_i \\ v'_i \\ w'_i \end{pmatrix} = R' \begin{pmatrix} x'_i \\ y'_i \\ -f \end{pmatrix} = \begin{pmatrix} x'_i \\ y'_i \\ -f \end{pmatrix}$$

and

$$\bar{p}''_i = \begin{pmatrix} u''_i \\ v''_i \\ w''_i \end{pmatrix} = R'' \begin{pmatrix} x''_i \\ y''_i \\ -f \end{pmatrix}$$

The coplanarity equation may be written in determinant form with the components in the system of the base:

$$\begin{bmatrix} bx & by & bz \\ u'_i & v'_i & w'_i \\ u''_i & v''_i & w''_i \end{bmatrix} = 0 = \begin{bmatrix} bx & by & bz \\ x'_i & y'_i & -f \\ u''_i & v''_i & w''_i \end{bmatrix}$$

In order to obtain observation equations, linearization by differentiation and Taylor expansion becomes possible:

$$\begin{vmatrix} bx & by & bz \\ 1 & 0 & 0 \\ u''_i & v''_i & w''_i \end{vmatrix} v'_{x_i} + \begin{vmatrix} bx & by & bz \\ 0 & 1 & 0 \\ u''_i & v''_i & w''_i \end{vmatrix} v'_{y_i} + \begin{vmatrix} bx & by & bz \\ x'_i & y'_i & -f \\ 1 & 0 & 0 \end{vmatrix} v''_{x_i}$$

$$+ \begin{vmatrix} bx & by & bz \\ x'_i & y'_i & -f \\ 0 & 1 & 0 \end{vmatrix} v''_{y_i} + \begin{vmatrix} bx & by & bz \\ x'_i & y'_i & -f \\ \dfrac{\delta u''_i}{\delta\omega''} & \dfrac{\delta v''_i}{\delta\omega''} & \dfrac{\delta w''_i}{\delta\omega''} \end{vmatrix} \delta\omega'' + \begin{vmatrix} bx & by & bz \\ x'_i & y'_i & -f \\ \dfrac{\delta u''_i}{\delta\varphi''} & \dfrac{\delta v''_i}{\delta\varphi''} & \dfrac{\delta w''_i}{\delta\varphi''} \end{vmatrix} \delta\varphi''$$

$$+ \begin{vmatrix} bx & by & bz \\ x'_i & y'_i & -f \\ \dfrac{\delta u''_i}{\delta\kappa''} & \dfrac{\delta v''_i}{\delta\kappa''} & \dfrac{\delta w''_i}{\delta\kappa''} \end{vmatrix} \delta\kappa'' + \begin{vmatrix} 0 & dby & dbz \\ x'_i & y'_i & -f \\ u''_i & v''_i & w''_i \end{vmatrix} - \begin{vmatrix} bx & by & bz \\ x'_i & y'_i & -f \\ u''_i & v''_i & w''_i \end{vmatrix}_0$$

The observation equations contain the observation errors in correlated form:

$$\underset{j,i}{Bv_{i,1}} = \underset{i,j}{A}\; \underset{j,i}{x} - \underset{j,l}{\ell_{i,1}}$$

For near vertical photography these correlations are, however, negligible, so that:

$$Bv = b \cdot f(v''_{yi} - v'_{yi}) = b \cdot f(py_i)$$

$v''_{yi} - v'_{yi}$ is the y-parallax, py_i, for which the error equations may be written as follows:

$$v'_{py} = \frac{1}{bf} \left\{ \begin{array}{l} (-fu''_i - x'_i w''_i)dby + (x'_i v''_i - y'_i u''_i)dbz + [bx(y'_i v''_i - fw'_i) \\ dbz - by(x'_i v''_i) - bz(x'_i w''_i)]d\omega + \\ [by(fw''_i - x'_i u''_i) + bz(-fw''_i)]d\varphi'' + [bx(fu''_i) + by(fv''_i) + bz \\ (x'_i u''_i + y'_i v''_i)]d\kappa'' \end{array} \right.$$
$$\left. - \left\{ py_{\mathrm{measured}} - \frac{1}{bf}[bx(y'_i w''_i + fv''_i) - by(x'_i w''_i + fu''_i) + bz \atop (y'_i v''_i + y'_i w''_i)] \right\} \right\}$$

These error equations may be set up for the six von Gruber points, in which the y-parallaxes have been measured.

$$x_{5,1} = \left(\underset{5,6\ 6,5}{A^T A}\right)^{-1} \cdot \underset{5,6\ 6,1}{A^T \ell}$$

will yield in the orientation corrections db_y, db_z, $d\omega''$, $d\varphi''$ and $d\kappa''$ for the new photo. With these model coordinates of the orientation, points may be calculated by a space intersection with $bx = 1$ or with an arbitrary value.

It is possible to link the next stereo model to the first in this manner by a scale transfer, as shown in Figure 3.35.

Absolute orientation

The next task is to orient the model with respect to the ground by a simple seven-parameter transformation. This process is called 'absolute orientation' (see Figure 3.36).

The model coordinate system, x_i, y_i, z_i, has its origin at M and the ground coordinate system, X_i, Y_i, Z_i at G, G, M and Pi form a triangle where the vectors ξ_o, $\lambda \cdot \vec{x}_i$ and ξ_i may be added to form:

$$\xi_i = \xi_o + \lambda \cdot T \cdot \vec{x}_i$$

$$\begin{pmatrix} X_i \\ Y_i \\ Z_i \end{pmatrix} = \begin{pmatrix} X_o \\ Y_o \\ Z_o \end{pmatrix} + \lambda \cdot T \begin{pmatrix} x_i \\ y_i \\ z_i \end{pmatrix}$$

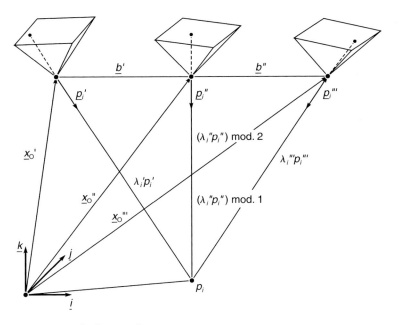

Figure 3.35 Scale transfer.

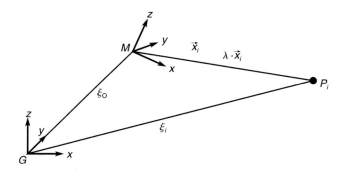

Figure 3.36 Absolute orientation.

T is a rotational matrix with the rotations Ω, Φ and K for the absolute orientation. X_o, Y_o and Z_o are the shift parameters to refer the coordinates to the origin of the ground system, and λ is the scale factor for the model.

Again error equations may be set up for the observed model coordinates. Their correlations may be neglected in absolute orientation.

$$\begin{pmatrix} v_x \\ v_y \\ v_{z_i} \end{pmatrix} = \frac{1}{\lambda} T^T \begin{pmatrix} X_i - X_o \\ Y_i - Y_o \\ Z_i - Z_o \end{pmatrix} - \begin{pmatrix} x_i \\ y_i \\ z_i \end{pmatrix}_{observed}$$

In linearized form, the equations become:

$$\begin{pmatrix} v_x \\ v_y \\ v_{z_i} \end{pmatrix} = \begin{pmatrix} \frac{\delta x_i}{\delta \lambda}d\lambda + \frac{\delta x_i}{\delta X_o}dX_o + \frac{\delta x_i}{\delta Y_o}dY_o + \frac{\delta x_i}{\delta Z_o}dZ_o + \frac{\delta x_i}{\delta \Omega}d\Omega + \frac{\delta x_i}{\delta \Phi}d\Phi + \frac{\delta x_i}{\delta K}dK \\ \frac{\delta y_i}{\delta \lambda}d\lambda + \frac{\delta y_i}{\delta X_o}dX_o + \frac{\delta y_i}{\delta Y_o}dY_o + \frac{\delta y_i}{\delta Z_o}dZ_o + \frac{\delta y_i}{\delta \Omega}d\Omega + \frac{\delta y_i}{\delta \Phi}d\Phi + \frac{\delta y_i}{\delta K}dK \\ \frac{\delta z_i}{\delta \lambda}d\lambda + \frac{\delta z_i}{\delta X_o}dX_o + \frac{\delta z_i}{\delta Y_o}dY_o + \frac{\delta z_i}{\delta Z_o}dZ_o + \frac{\delta z_i}{\delta \Omega}d\Omega + \frac{\delta z_i}{\delta \Phi}d\Phi + \frac{\delta z_i}{\delta K}dK \end{pmatrix}$$

$$- \begin{pmatrix} x_i - (x_i)_0 \\ (y_i - (y_i)_0 \\ (z_i - (z_i)_0 \end{pmatrix}$$

For near vertical photography, the relations can be simplified to:

$$v_x = \lambda \cdot x_i d\lambda + \lambda \cdot z_i d\phi + \lambda \cdot y_i d\kappa + dX_o - \lambda \cdot x_i + (X_0)_0$$

$$v_y = \lambda \cdot y_i d\lambda + \lambda \cdot z_i d\Omega - \lambda \cdot x_i d\kappa + dY_o - \lambda \cdot y_i + (Y_0)_0$$

$$v_z = \lambda \cdot z_i d\lambda - \lambda \cdot y_i d\Omega - \lambda \cdot y_i d\phi + dZ_o - \lambda \cdot z_i + (Z_0)_0$$

This equation system may be solved by a least squares adjustment to obtain the unknowns:

$$\lambda = \lambda_0 + d\lambda, \ \Omega_0 + d\Omega, \ \Phi = \Phi_0 + d\Phi, \ K = K_0 + dK, \ X_o = (X_o)_o + dX_o,$$
$$Y_o = (Y_o) + dY_o$$

and $z_0 = (z_0)_0 + dZ_0$ to calculate the ground coordinates, X_i, Y_i and Z_i of the model.

For the solution of absolute orientation of a stereo model, three control points forming a control point triangle are required (see Figure 3.15).

The application of this absolute orientation solution has also made it possible to perform an adjustment of aerial triangulation blocks for stereo models. For this, it is necessary to form spatial triangles between the adjacent von Gruber points of two models and the common perspective centre of the exposure station. This solution was realized in F. Ackermann's PAT-M adjustment (see Figure 3.37).

Digital photogrammetric operations

The sequence of digital photogrammetric operations is described here with the help of the program package SIDIP (Simple Digital Photogrammetry), which originated at the University of Hannover. The core of the program is the bundle block adjustment (BLUH), programmed by K. Jacobsen of the University's Institute for Photogrammetry and GeoInformation. It was linked with semi-automatic photo measurement packages (DPLX by B.

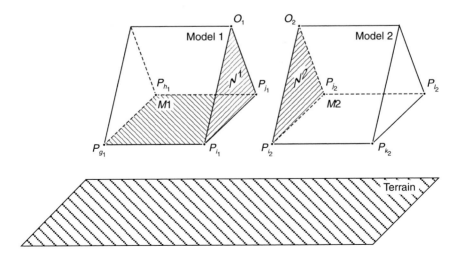

Figure 3.37 Model aerial triangulation.

Pollak and DPCOR of the Institute). The DEM, orthophoto generation and the GIS interface was provided by W. Linder of the Geographic Institute of the University of Düsseldorf (LISA-FOTO and LISA-Basic).

SIDIP

The program package SIDIP is an example of a basic tool in digital photogrammetry. It does not possess the customized convenience for automated operations in a full production process, such as is contained in Z/I Imaging's Image Station 2001 or in LH-System's SOCET-SET, but it contains all of its basic elements. Alongside an explanation of SIDIP, references will be made here to these more powerful workstation solutions. SIDIP operates on a minimal configuration of a standard PC with 64 Mb RAM and 5 GB of storage in a standard Windows environment. It consists of five program parts shown in Figure 3.38.

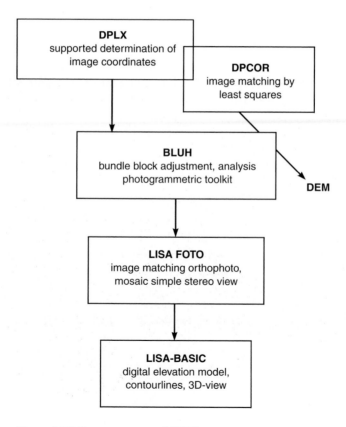

Figure 3.38 Program parts of SIDIP.

DPLX

DPLX is a program for the semiautomatic measurement of points in images. Due to limitations in display storage of a full photogrammetric image, overview images are created in the form of image pyramids, which display the average of a specific number of pixel grey values (e.g. 4×4) as one pixel on the screen. A frame in the overview image can be selected to display a selected portion at full resolution. At the original image resolution, points may be selected for measurement. To more precisely select the pointing zoom images (e.g. 300 per cent) may be created. At a click of a mouse, the measured image point will be marked with a cross and its set or automatically advanced point number (see Figure 3.39).

The movement of the measuring mark in the original image by the mouse is programmed for a fixed display of the image part.

More powerful commercial workstations can display the entire images directly, have a continuous zooming possibility and permit roaming. During the roaming operation the measuring mark remains centred and the image is moved in relation to it. The measurement begins with the four or eight fiducial marks of the image. They permit the location of its principal point (see Figure 3.40).

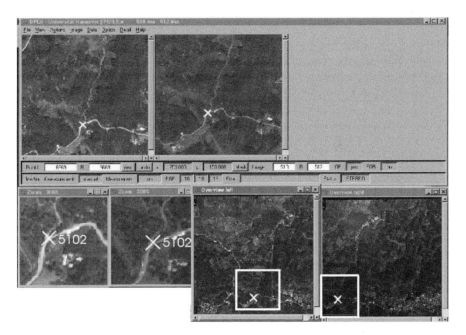

Figure 3.39 DPLX screen.

Source: Image courtesy of Institute for Photogrammetry and GeoInformation, University of Hannover.

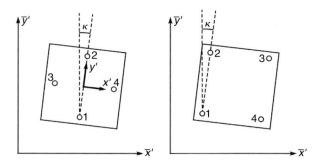

Figure 3.40 Measurement of fiducial marks.

Source: Image courtesy of Institute for Photogrammetry and GeoInformation, University of Hannover.

The measurement is made in pixel coordinates $\vec{x}'_i$, $\vec{y}'_i$. These can be transformed into image coordinates x'_i, y'_i by the following relations:

$$\begin{pmatrix} x'_i \\ y'_i \end{pmatrix} = \begin{pmatrix} \cos \kappa - \sin \kappa \\ \sin \kappa - \cos \kappa \end{pmatrix} \begin{pmatrix} \vec{x}'_i - \vec{x}'_o \\ \vec{y}'_i - \vec{y}'_o \end{pmatrix}$$

in which, for four fiducial marks 1 to 4, the principal point has the pixel coordinates:

$$\vec{x}'_o = \frac{\vec{x}'_1 + \vec{x}'_2 + \vec{x}'_3 - \vec{x}'_4}{4}$$

$$\vec{y}'_o = \frac{\vec{y}'_1 + \vec{y}'_2 + \vec{y}'_3 - \vec{y}'_4}{4}$$

or the transformation is carried out by an affine model:

$$x'_i = a_1 + a_2 \vec{x}'_i + a_3 \vec{y}'_i$$

$$y'_i = a_4 + a_5 \vec{x}'_i + a_6 \vec{y}'_i$$

and

$$tg\kappa = \frac{\vec{x}'_2 + \vec{x}'_1}{\vec{y}'_2 + \vec{y}'_1}$$

This process is called 'interior orientation'. For four or more fiducial marks, it may be subjected to a least squares adjustment with indication of the residuals.

The fiducial marks have round images in the photo. This permits them to be accurately measured automatically by an ellipse operator applicable to circular or ellipse-shaped targets. In the pixel grid covering the target, the grey level edges of the target are identified for each horizontal and vertical line of that grid. Their position is halved for all horizontal and vertical lines. These mid positions of the lines can be linked by two straight lines, the intersection of which results in the coordinates of the fiducial mark. DPLX includes not only the automatic measurement of fiducial marks, but also the automatic measurement of signalized points or of symmetrical manholes (see Figure 3.41).

More advanced workstations have the added possibility of performing the entire interior orientation process automatically.

For measurements on fiducials and image points, an alphanumeric display permits the display of measured points in both photos with a listing of point number, pixel coordinates $\tilde{x}'_i$, $\tilde{y}'_i$, and image coordinates x'_i, y'_i. The listing shows the measured fiducial marks and the selected terrain points measured as control points and terrain points visible at least in two images or as transfer points visible in three images along a flight strip, or in six images across two strips.

DPCOR

The module DPCOR is integrated into DPLX for automatic image matching, which is based on a region growing method starting from seed points. DPCOR uses an image correlation for the determination of the approximate positions of corresponding points, which will be improved by least squares matching. Figure 3.42 shows the screen display of the equivalent ERDAS Imaging/Stereoanalyst software.

BLUH

The image coordinates computed by the interior orientation process are

Figure 3.41 Automatic measurement of fiducial marks.

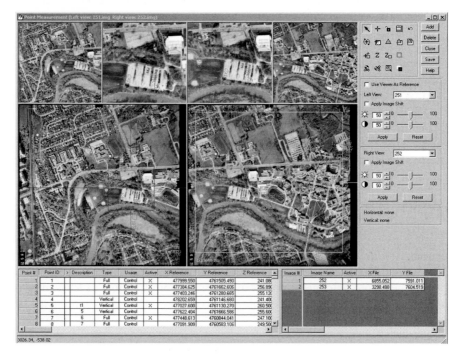

Figure 3.42 Point measurement on the Erdas system.

Source: Stereo imagery of Los Angeles, CA; data provider: HJW Inc., software tool: Imagine OrthoBASE ® Erdas; illustration provider: Erdas Inc., the Geographic Imaging unit within Leica Geosystems GIS & Mapping Division, Atlanta, GA.

passed on into the bundle block adjustment program BLUH, consisting of more than fifty modules, the most important functions of which are:

- preparation of photo coordinates; this permits the changing of point numbers or the numbers of groups of points, the sorting of the observations and the correction of photo coordinates using calibration values for lens distortion, earth curvature, refraction and film deformations by transformation to the measured fiducial marks. The residuals of the fiducial mark measurements are indicated with a check of acceptability.
- calculation of approximate photo orientations, the search for errors in the point measurements and the sorting of observations to conform to a reduced bandwidth for the normal equations. With the aid of identical point numbers in the photos of a strip object coordinates are calculated by two-dimensional transformation. Thereafter adjacent strips treated in the same manner are linked in two dimensions. All compu-

tations are checked by data snooping to detect errors, which are indicated and eliminated.

- the setting of options for output files, the use of a particular choice of additional parameters, of error limits and weights for photo and ground coordinate measurements, the setting of robust estimator parameters for the adjustment and the option to utilize additional measurements for in-flight GPS exposure stations and IMU exposure directions with appropriate weights.

- the execution of the least squares adjustment solution to determine the parameters of orientation, the measured points and the additional parameters determined.

- the output in numerical form consists of lists of the observation errors, the achieved σ_o and the object coordinates of all points measured as well as the image orientations.

- the output in graphical form shows the images and the discrepancy vectors with respect to the ground control points used. It is also possible to show the effect of additional parameters on the image coordinates (see Figure 3.43).

  The required geodetic coordinate conversions, in the case of high altitude and space images, are also included as an option.

- for scanner images, the standard geometric model using the perspective

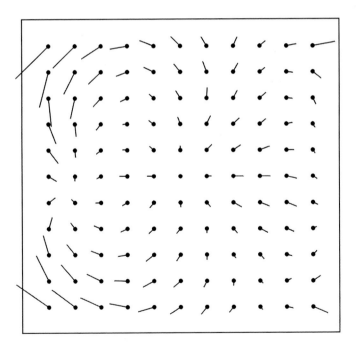

Figure 3.43 Image coordinate corrections determined by additional
           parameters.

collinearity equations, needs to be replaced as an option. Instead of the collinearity equations a separate scanner geometry is substituted. For a scanner, the image projection centre x'_o moves forward from a beginning position, x'_{o_B}, to an end position, x'_{o_E}, in the image. The in-between projection centre, x'_{o_i}, at distance b_i, can be interpolated along the base, b_{BE}:

$$x'_{o_i} = x'_{o_A} + \frac{b_i}{b_{BE}}(x'_{o_E} - x'_{o_B})$$

i represents the number of the scan line as a function of time.

An image point along the scan line, i, will therefore have the coordinates:

$$x'_i = (0,\, y',\, -f)$$

The imaged object point, x_i, along the line can thus be determined from:

$$X_i = \lambda \cdot R \cdot x'_i + x'_{o_i}$$

As a first approximation, the orientation matrix, R, and its parameters, φ, ω, κ, can be taken as constant for a satellite scanner image. A number of control points can assure that the orientation changes during the platform motion are modelled as additional parameters.

For an aircraft scanner image, φ_i, ω_i and κ_i, along the flight strip must be known as a function of time, t_i, through an IMU.

LISA-FOTO

LISA-FOTO performs the tasks of a digital photogrammetric workstation. Figure 3.44 shows the user interface of the program.

The digitally stored photos used in DPLX and the photo orientations calculated in BLUH are imported into LISA-FOTO. This program has the following essential parts:

- stereo measurement.
- stereo correlation.
- ortho image generation.

Through the known image orientations from BLUH, the stereo model can be displayed in epipolar mode with no y-parallaxes present. Stereo viewing can be achieved in two simple modes: viewing of the images by a stereo-scope or by anaglyphs.

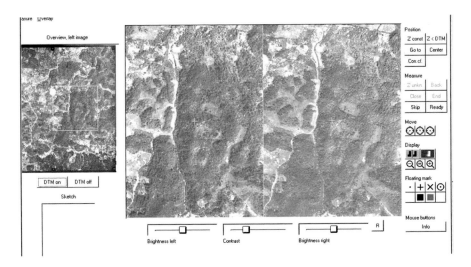

Figure 3.44 LISA-FOTO user interface.

Source: Image courtesy of Institute for Photogrammetry and GeoInformation, University of Hannover.

The stereo measurement, like in DPLX, is initiated by selecting a model part on an overview image. Unlike DPLX, the program has roaming facilities in fast, intermediate and slow motion, moving the images with respect to a fixed floating mark of changeable shape and colour. The floating point can be kept at a constant height, or it can follow the elevation of a stored geocoded DEM. This is ideal for extracting vector information from the model since no changes of the floating mark are required during the plotting information.

Image matching

Image matching or stereo correlation has a long history. In 1957, Gilbert Hobrough patented an electronic image correlation. These initiatives sparked a number of further industrial developments by Raytheon-Wild (Stereomat B8), Bunker-Ramo-Wooldridge (Unamace), Bendix and Gestalt to permit the automatic generation of digital elevation models by electronic image correlation. Through the present availability of high speed, large storage, digital computing these attempts have become obsolete.

Digital image matching may be performed in different ways. The simplest method is the comparison of the cross correlation coefficient between two images to be matched. A pattern matrix of a limited dimension (e.g. 17×17 pixels) of one image with grey values, d'_{ij}, can be compared with an equivalent size matrix of a second image with the grey values, d''_{ij}. For these two images, the variances σ', σ'' and the covariance r_{ij} may be calculated:

$$\sigma' = \sqrt{\frac{1}{n-1}\Sigma(d'_{ij}-d')^2} \text{ with } d' = \frac{1}{n}\sum_{1}^{n}d'_{ij}$$

$$\sigma'' = \sqrt{\frac{1}{n-1}\Sigma(d''_{ij}-d'')^2} \text{ with } d'' = \frac{1}{n}\sum_{1}^{n}d''_{ij}$$

$$\operatorname{cov}r_{ij} = \frac{1}{n}\Sigma d'_{ij}\cdot d''_{ij}$$

the cross correlation coefficient, r_{ij}, becomes:

$$r_{ij} = \frac{\operatorname{cov}r_{ij}}{\sigma'\cdot\sigma''}$$

This calculation is repeated within the search area larger than the kernel of the pattern matrix and the first search matrix by shifting the search matrix from left to right and from up to down. The result will be a matrix of cross correlation coefficients and their largest value will correspond to the x and y shift of the best match. The sequence of matches in x and then in y direction should result in a curve for the correlation coefficient with a maximum value. The search configuration is shown in Figure 3.45.

If the maximal correlation coefficient is greater than 0.7, a good match has been found. Depending on the contrast, the illumination differences or the slope of the terrain, a significantly lower correlation coefficient may be obtained or the correlation might be lost.

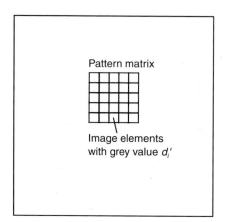
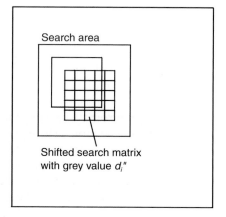

Figure 3.45 Calculation of cross correlation coefficients.

One strategy to find the required approximate positions and to avoid loss of correlation is to work with image pyramids averaging 2×2, 4×4, 8×8 and 16×16 pixels into one density value. The correlation may start with the coarsest pyramid overview image. This will give approximations to start correlation at the next finer pyramid until the maximum correlation is reached at the highest pyramid level (see Figure 3.46).

Another strategy is to apply least squares matching. The aim of the method is to minimize the square sum of grey level differences between the pattern matrix and the geometrically transformed search matrix. In a first approximation, the transformation can be affine, or at a more refined level it can be projective. In image correlation, each density value, $d'_i(x, y)$, of the pattern matrix should correspond to the density value, $d''_i(x, y)$, of the search matrix so that:

$$d'_i(x, y) - v_i(x, y) = d''_i(x, y)$$

Assuming an affine deformation of the search matrix, the equation may be expanded:

$$d'_i(x, y) - v_i(x, y) = r_o + r_1 \cdot d''_i(x', y')$$

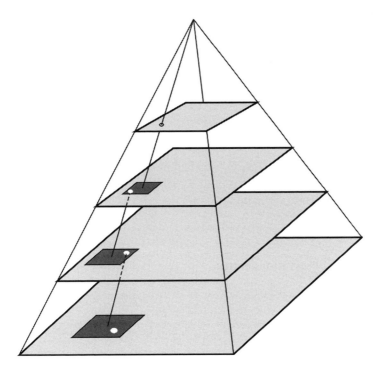

Figure 3.46 Image pyramids.

with

$$x' = a_o + a_1 x + a_2 y$$

$$y' = b_o + b_1 x + b_2 y$$

The linearization of the equation results in

$$d'_i(x, y) - v_i(x, y) = d''_i(x, y) - d''_o(x, y)$$

$$+ \frac{\delta d''(x, y)}{\delta x} da_o$$

$$+ \frac{\delta d''(x, y)}{\delta x} da_1$$

$$+ \frac{\delta d''(x, y)}{\delta x} da_2$$

$$+ \frac{\delta d''(x, y)}{\delta y} db_o$$

$$+ \frac{\delta d''(x, y)}{\delta y} db_1$$

$$+ \frac{\delta d''(x, y)}{\delta y} db_2$$

The approximation value, $d''_o(x, y)$, is calculated with the assumptions:

$$a_o = a_2 = b_o = b_2 = r_o = 0, \text{ and}$$

$$a_1 = b_x = r_1 = 1$$

This leads to the observation matrix:

$$\ell_{n,1} + v_{n,1} = A_{n,u} X_{u,1}$$

$X_{u,1}^T = da_o, da_1, da_2, db_o, db_1, db_2, r_o, r_1$ can be obtained from a least squares solution

$$x = (A^T A)^{-1} A^T \ell$$

with

$$\sigma_o = \sqrt{\frac{v^T v}{n-1}}$$

In practice, correlation can start at the coarsest pyramid by calculation of the cross correlation coefficient. This is repeated until the finest pyramid. Following that, least squares matching is applied.

The correlation algorithms discussed thus far have to be executed for every pixel of the overlapping image matrix in which the pattern matrix (and the search matrix) is systematically moved around in horizontal and vertical direction. This leads to a time-consuming computation which, depending on pixel size, can take from minutes to hours on a small PC.

To economize on computation time, a selection of the measurement effort can be made: it is possible to correlate only in a regular point grid with the intent to interpolate elevation values later.

To narrow the extent of the search area, a regular object grid is chosen for which the location of the pattern matrix centre and the search matrix centre is calculated at different plausible elevations. For these locations the elevation corresponding to the highest cross correlation coefficient is chosen as the correlation result. This procedure is utilized in LISA-FOTO. DPCOR can attempt to improve the obtained correlation result applying least squares matching.

Another possibility is the selection of points, where correlation is to be performed by an interest operator, which selects from all image pixels those which are most suitable for an accurate correlation because of their symmetrical and large enough contrasts.

One of these operators is the Moravec operator. It calculates the mean square gradient, V_1, V_2, V_3, V_4, of a pixel window, $n \times n$, in all four main directions:

$$V_1 = \frac{1}{n(n-1)} \Sigma\Sigma[d(i, j) - d(i, j+1)]^2$$

$$V_2 = \frac{1}{n(n-1)} \Sigma\Sigma[d(i, j) - d(i+1, j)]^2$$

$$V_3 = \frac{1}{n(n-1)^2} \Sigma\Sigma[d(i, j) - d(i, j+1)]^2$$

$$V_4 = \frac{1}{n(n-1)^2} \Sigma\Sigma[d(i, j) - d(i, j+1)]^2$$

The objective is to find a min (V_1, V_2, V_3, V_4).

Another suitable operator is the Förstner operator. It is based on the model that the transformed grey level vicinity of a point, $d'(x, y)$, is a function of the original image, $d(x + x_o, y + y_o)$. This gives rise to an error equation:

$$d'(x, y) = d(x + x_o, y + y_o) + (x, y)$$

After linearization, setting the starting values, x_o, $y_o = 0$, this equation becomes:

$$dd(x, y) - v(x, y) = \frac{\partial d}{\partial x} x_o + \frac{\partial d}{\partial y} y_o$$

with

$$dd(x, y) = d'(x, y) + d(x, y)$$

This leads to a normal equation matrix, N:

$$\begin{pmatrix} \Sigma\left(\frac{\partial g}{\partial x}\right)^2 & \Sigma\left(\frac{\partial g}{\partial x}\frac{\partial g}{\partial y}\right) \\ \Sigma\left(\frac{\partial g}{\partial y}\frac{\partial g}{\partial x}\right) & \Sigma\left(\frac{\partial g}{\partial y}\right)^2 \end{pmatrix} \begin{pmatrix} x_o \\ y_o \end{pmatrix} = \begin{pmatrix} \frac{\partial g}{\partial x} \cdot dd(x, y) \\ \frac{\partial g}{\partial y} \cdot dd(x, y) \end{pmatrix}$$

or

$$A^T A x = A^T(dd(x, y)) = Nx$$

This matrix, N, can be rotated to obtain an uncorrelated (diagonal) eigenvalue matrix, which permits the calculation of two characteristic values, w and q, for an error ellipse, so that

$$w = \frac{1}{\lambda_1 + \lambda_2}$$

and

$$q = 1 - \left(\frac{\lambda_1 - \lambda_2}{\lambda_1 + \lambda_2}\right)^2$$

w represents the size of the ellipse and q its roundness.

A point selected by the Förstner operator should have a minimum value, W_{min}, of:

$$W_{min} = 0.5 \text{ to } 1.5(W_{\text{average for the entire image}})$$

and

$$q_{min} = 0.5 \text{ to } 7.5$$

The choice of interest operator points significantly reduces the image matching effort for the entire model. Other approaches for image matching have been suggested using extraction of line features in both images and in subsequent relational matching of the line features.

Image matching is usually done in image space. A resampling of the correlated pixel result into object space by resampling, according to Chapter 2 (pages 84–86) is then required. If the chosen correlation algorithm works in object space, it will in practice only provide a limited number of object points correlated, and the creation of an object pixel based digital elevation matrix is also needed.

Digital elevation models

The result of image matching is the creation of elevation data, z_i, at a specified object location, x_i, y_i. Depending on the correlation algorithm, these are located either in a regular grid (as in LISA-FOTO) or spaced in an irregular distribution (as by use of interest operators). In both cases an interpolation of these heights to a digital elevation matrix related to output pixels in the object system is required.

The interpolation can be done using the already determined height points in a stochastic field (see Figure 3.47).

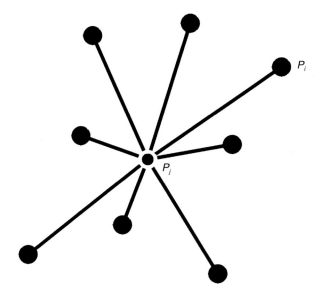

Figure 3.47 Interpolation in a stochastic field.

P_i are the points with known elevations, z_i, in the vicinity of point, P_j, where z_j is to be interpolated. Interpolation can be done using a weight function, P_{ij}, which is proportional to the distance between P_i and P_j:

$$p_{ij} = \frac{1}{\sqrt{(x_j - x_i)^2 + (y_j - y_i)^2}}$$

$$z_j = \frac{\Sigma p_{ij} z_i}{\Sigma p_{ij}}$$

As weight function the square of the distance may also be used:

$$p_{ij} = \frac{1}{(x_j - x_i)^2 + (y_j - y_i)^2}$$

To consider a directional dependence, the location z_j can be expressed on an oblique plane with:

$$z_j = a_o + a_1 x_j + a_2 y_j$$

The coefficients, a_o, a_1, a_2, can be calculated as an adjustment problem from the nearest points, P_i, with the weights, p_{ij}.

The interpolation model can finally be expanded into least squares interpolation, which determines the distance dependent covariance function from all observed points and uses it as a weight to obtain z_j. Its advantage is to permit smoothing of the observation (K. Kraus). Another approach is the interpolation on the basis of finite elements (H. Ebner). Finally, it becomes possible to interpolate height on the basis of triangulated irregular networks (TIN) (see Figure 3.48).

Between all points at which elevations were measured, distances can be calculated. Three of these distances can be combined into a particular triangle. The triangle with the smallest area is finally chosen for the TIN, which consists of an assembly of smallest triangles generated.

For the triangle $P_1 P_2 P_3$ with its coordinates $(x_1 y_1, x_2 y_2, x_3 y_3)$ and heights (z_1, z_2, z_3) in which interpolation is to take place, z_i is expressed on the plane formed by the edges of the triangle:

$$z_i = a_o + a_1 x_i + a_2 y_i$$

The coefficients, a_o, a_1, a_2, can be calculated from the three equations for that triangle:

$$\begin{pmatrix} 1 & x_1 & y_2 \\ 0 & x_2 & y_2 \\ 0 & x_3 & y_3 \end{pmatrix} \begin{pmatrix} a_o \\ a_1 \\ a_2 \end{pmatrix} = \begin{pmatrix} z_1 \\ z_2 \\ z_3 \end{pmatrix}$$

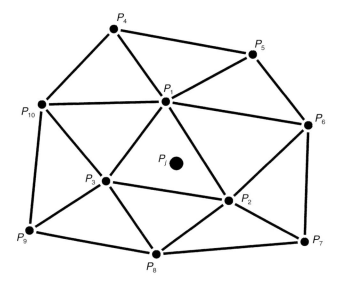

Figure 3.48 TIN interpolation.

If modelled in this way, adjacent triangles will not represent a smooth surface, but they will be separated by sharp edges.

To permit a smooth transition between triangles, the triangle surface may be expanded into a polynomial:

$$z_i = a_o + a_1x_i + a_2x_i^2 + a_3x_i^3 + a_4y_i + a_5y_i^2 + a_6y_i^3 + a_7x_iy_i + a_8x_i^2y_i + a_9x_iy_i^2$$

To solve for the coefficients of this polynomial, not only the heights, z_1, z_2, z_3, are available as observations, but also the slopes and curvatures in x and y of the adjacent triangles formed with P_4 to P_{10}.

The following eighteen observation equations can be used for the triangle.

For the elevations:

$$z_1 = a_o + a_1x_1 + \ldots a_9x_1y_1^2$$

$$z_2 = a_o + a_1x_2 + \ldots a_9x_2y_2^2$$

$$z_3 = a_o + a_1x_3 + \ldots a_9x_3y_3^2$$

For the slope:

$$\frac{\partial z_i}{\partial x_1} = a_1 + 2a_2x_1 + 3a_3x_1^2 + a_7y_1 + 2a_8x_1y_1 + a_9y_1^2$$

$$\frac{\partial z_i}{\partial y_1} = a_4 + 2a_5y_1 + 3a_6y_1^2 + a_7x_1 + a_8x^2 + 2a_9x_1y_1$$

similarly for

$$\frac{\partial z_2}{\partial x_2}, \frac{\partial z_2}{\partial y_2}, \frac{\partial z_3}{\partial x_3}, \frac{\partial z_3}{\partial y_3}$$

For the curvature:

$$\frac{\partial^2 z_1}{\partial x_1^2} = 2a_2 + 6a_3x_1 + 2a_8y_1$$

$$\frac{\partial^2 z_1}{\partial y_1^2} = 2a_5 + 6a_6y_1 + 2a_9x_1$$

$$\frac{\partial^2 z_1^2}{\partial x_1\partial_1} = a_1a_4 + 2a_2a_4x_1 + 2a_1a_5y$$

similarly for

$$\frac{\partial^2 z_2}{\partial x_2^2}, \frac{\partial^2 z_2}{\partial y_2^2}, \ldots \frac{\partial^2 z_3}{\partial x_3\partial y_3}$$

The other triangles will have similar observation equations. They can be solved in a simultaneous least squares adjustment for the ten or more coefficients selected.

The advantage of the TIN modelling is that, along the lines of a particular triangle, natural discontinuities in the form of break lines can be considered in the interpolation process. In case of a break line, the equations for slope and curvature between the triangles that have a discontinuity are simply omitted.

At the end of the interpolation process each output pixel, x_i, y_i, in object coordinates will have its corresponding z_i value. Figures 3.49, 3.50 and 3.51 show DEM processing procedures.

Orthoimage generation

With x_i, y_i, z_i known for each output pixel, the calculation of an orthoimage now becomes a simple matter. For the grey level transfer from one of the images forming a stereo model, the collinearity equations are used to calculate the corresponding, x_i' y_i' position. From there the grey value, d_i', is transferred to the output pixel as d_i, either by nearest neighbour

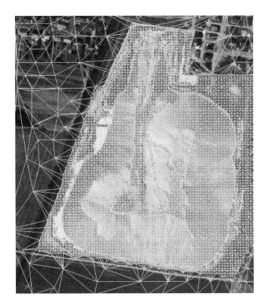

Figure 3.49 TIN structure of a DEM.

Source: Image courtesy of Institute for Photogrammetry and GeoInformation, University of Hannover.

Figure 3.50 Interpolation of contours from the TIN.

Source: Image courtesy of Institute for Photogrammetry and GeoInformation, University of Hannover.

Figure 3.51 Excessive slope calculation.

Source: Image courtesy of Institute for Photogrammetry and GeoInformation, University of Hannover.

assignment, by bilinear interpolation or by cubic convolution (see pages 84–86).

In the collinearity model, all image corrections determined from additional parameters $\Delta x_i'$ and $\Delta y_i'$ need to be included:

$$x_i' = -f \frac{r_{11}(x_i - x_o') + r_{21}(y_i - y_o') + r_{31}(z_i - z_o')}{r_{13}(x_i - x_o') + r_{23}(y_i - y_o') + r_{33}(z_i - z_o')} + \Delta x_i'$$

$$y_i' = -f \frac{r_{12}(x_i - x_o') + r_{22}(y_i - y_o') + a_{32}(z_i - z_o')}{r_{13}(x_i - x_o') + a_{23}(y_i - y_o') + r_{33}(z_i - z_o')} + \Delta y_i'$$

with

$$\Delta x_i' = a_1 x_i'(r_i' - r_o') + \ldots - a_9 x_i'$$

$$\Delta y_i' = a_1 y_i'(r_i' - r_o') + \ldots - a_{10} x_i'$$

(see page 152).

In the case of scanner images, the collinearity model needs to be replaced by the sensor model (see page 164). Otherwise the procedure is the same.

Adjacent orthophotos as part of different stereo models will have tone differences. To eliminate these, overlapping orthophoto areas may be created for which the orthophotos will be calculated from two photographs. The grey values from both of these object pixels may be taken as a mean. In this way ortho-mosaicking is accomplished.

The geocoded orthophoto coverage produced in this manner may be used directly for input and overlays in geographic information systems. The combination of digital orthophotos with vector information is possible by superimposition of raster and vector information. On the other hand, on-screen digitization with GIS software (MicroStation, Autocad, Geomedia, Arc-View or ArcInfo, etc.) becomes possible. Figure 3.52 shows an example of the superimposition of GIS information.

In the transfer of data, the format plays a role:

- raw data, in which the grey levels of an image are stored uncom-

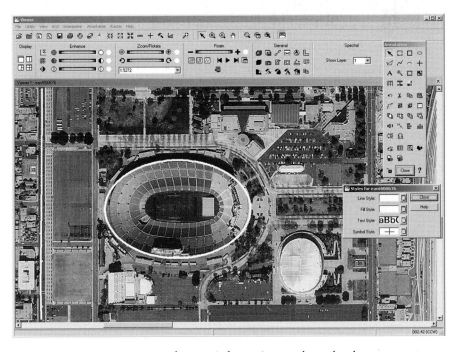

Figure 3.52 Superimposition of vector information on the orthophoto.

Source: Stereo imagery of Los Angeles, CA; data provider: HJW Inc., software tool: Imagine OrthoBASE ® Erdas; illustration provider: Erdas Inc., the Geographic Imaging unit within Leica Geosystems GIS & Mapping Division, Atlanta, GA.

pressed in a binary file, are not conducive to facilitate the use of the data. For colour images there is the possibility of storing the pixels representing the three colours as either pixel-interleaved (three pixels together), line-interleaved or band-interleaved.

- the most common format is the TIFF data format, which contains a directory indicating image size, colour dimension and image resolution. GEOTIFF includes data on the geocoding parameters.
- the bitmap (BMP) format is suitable for 24-bit black and white and colour images only for use in Microsoft Windows.
- the GIF format makes lossless data compression possible, however, at the expense of only 8 bits per pixel.
- the JPEG format has the best data compression capability. With JPEG, a matrix of 48×48 pixels is formed. The grey levels of the 48×48 pixel matrix, taking up much storage space, are not transferred as data but as a linear function, the coefficients of which are transferred. Thus the volume of data is significantly reduced depending on the chosen compression level, but this cannot be done without loss.

A possible by-product of orthoimages are stereo orthophotos or stereo partners. If a DEM has been created these are easily computable. Orthophotos present a vertical view of the terrain. For the stereo partner, one can choose a particular constant viewing angle for creating x-parallaxes, Δx:

$$\Delta x_i = -\Delta z_i \cdot tg\alpha = -k \cdot \Delta z_i \text{ (see Figure 3.53)}$$

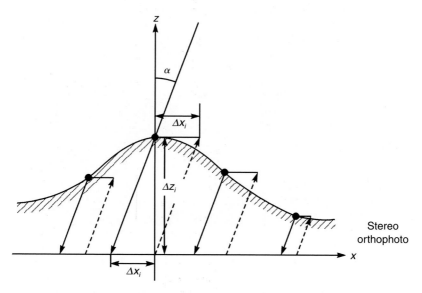

Figure 3.53 Stereo-orthophoto generation.

The stereo orthophoto can then be produced with the modified collinearity equations:

$$x_i' = -f \frac{r_{11}(x_i - k \cdot \Delta z_i - x_o') + r_{21}(y_i - y_o') + r_{31}(z_i - z_o')}{r_{13}(x_i - k \cdot \Delta z_i - x_o') + r_{23}(y_i - y_o') + r_{33}(z_i - z_o')}$$

$$y_i' = -f \frac{r_{12}(x_i - k \cdot \Delta z_i - x_o') + r_{22}(y_i - y_o') + r_{32}(z_i - z_o')}{r_{13}(x_i - k \cdot \Delta z_i - y_o') + r_{23}(y_i - y_o') + r_{33}(z_i - z_o')}$$

Orthophotos supply the correct geometry of a map and they have the advantage of offering visual interpretive capabilities. They do not, however, permit the display of object classes in the form of a GIS. Their use is therefore limited to acting as map substitutes for areas, where timely mapping is too costly (urban areas) or too slow, or where map classification does not permit sufficient object details (swamps, coastal areas, forest) to be shown.

At a large scale urban area orthophotos have particular problems: a digital elevation model attempts to record the elevation of ground points. An orthophoto produced with these elevations will, therefore, show radial displacements of buildings, trees or bridges. Research has therefore been aimed at developing a methodology for the generation of so-called 'true orthophotos', in which the displacements for the building tops or bridge levels have been removed case by case.

Image correlation provides height information on the top of objects. An urban or a forested scene, to which image matching was applied, therefore generates a digital surface model, rather than a digital terrain model.

SIDIP with the program RASCOR contains a filtering function by which elements not belonging to the surface may be ignored. This eliminates height measurements for building tops, and what remains is a DEM consisting of ground level heights.

ERDAS offers, with its packages Imagine and 3D GIS Stereo Analyst, the ability to vectorize, for example, a particular building in the ortho-photo and to differentially remove its displacements in the orthophoto (see Figure 3.54).

The semantic modelling group of W. Förstner at the Institute for Photogrammetry of the University of Bonn has been successful in the 3D modelling of buildings. Edge matching is carried out for buildings, fitting the derived form to a particular building type, allowing the transformation of that building into an orthogonal view.

A. Gruen of the ETH Zürich and D. Fritsch of the University of Stuttgart and their groups use this strategy to derive 3D city models in which the façades contained in the images are rectified into the 3D scene.

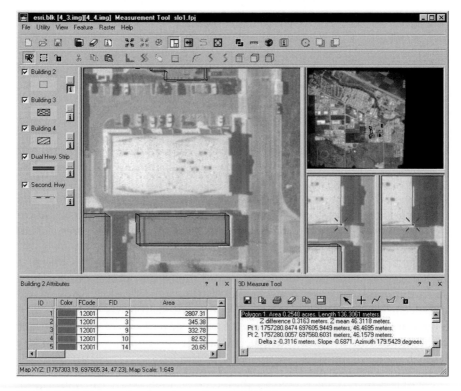

Figure 3.54 'True orthophoto' creation of ERDAS.

Source: Stereo imagery of Los Angeles, CA; data provider: HJW Inc., software tool: Imagine OrthoBASE ® Erdas; illustration provider: Erdas Inc., the Geographic Imaging unit within Leica Geosystems GIS & Mapping Division, Atlanta, GA.

LISA-Basic

LISA-Basic is a general digital elevation model program supported by break lines. Initially elevation zones are assigned grey levels. This raster DEM can be displayed as a raster black and white image or as a colour-coded image for the zones chosen. It is also possible to interpolate contours, to display profiles and to create oblique wire frame displays from these (see Figure 3.55).

Raster line data for contours may be converted into vector data for export in plot files in BMP, PCX or DXF format. Finally, the orthophoto may be draped over the oblique view of a DEM. The viewing position and the viewing direction can be arbitrarily chosen for the DEM. Then the ray between each DEM pixel and the viewing point is calculated and projected onto an image perpendicular to the viewing direction. The orthophoto

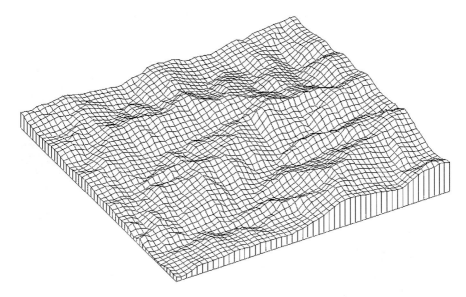

Figure 3.55 Wire frame display of a DEM.

Source: Image courtesy of Institute for Photogrammetry and GeoInformation, University of Hannover.

grey level at the DEM pixel is transferred to the projected image. A resampling of this image is executed to obtain the oblique view (see Figure 3.56).

In order to omit hidden areas, the pixel-by-pixel calculation is carried out starting with the longest distance between DEM pixel and viewing point. For this ray, the grey level is transferred first. Then the calculation is

Figure 3.56 Draped orthophoto projected onto a DEM.

Source: Image courtesy of Institute for Photogrammetry and GeoInformation, University of Hannover; map courtesy LGN, Hannover.

repeated for the next shorter direction, and so forth. In doing so the previous grey levels are overwritten, so that the hidden pixel grey values will not be visible in the oblique view.

LISA-Basic also has the capability to deduce secondary products of the DEM such as:

- slope images, for which the slopes for a pixel are calculated between the adjacent eight pixels. The steepness of the slope, separated in different levels, can then be displayed in grey shades or in colour code.
- aspect images, for which the incident solar illumination at a certain hour (noon), latitude and day of the year (declination) directed to a DEM pixel, determines an illumination vector. For an object point, a surface normal may be calculated from the adjacent elevation pixels. It is perpendicular to a plane expressed by x and y parameters. The angular difference between illumination vector and the vector of the surface normal may again be expressed in grey values or in colour code for different magnitude levels of that spatial angle.
- It is furthermore possible to compare two different DEMs for the purpose of error analysis or for change detection and to display the differences in an appropriate form with the described possibilities.

4 Geographic information systems

Introduction

A geographic information system (GIS), in a narrow definition, is a computer system for the input, manipulation, storage and output of digital spatial data. In a more broad definition it is a digital system for the acquisition, management, analysis and visualization of spatial data for the purposes of planning, administering and monitoring the natural and socio-economic environment. It represents a digital model of geography in its widest sense (see Figure 4.1).

In the narrow sense, a GIS consists of a system for data input in vector form, in raster form and in alphanumeric form, a CPU containing the programs for data processing, data storage and data analysis and of facilities for visualization and hard copy output of the data. In a broad sense, a GIS

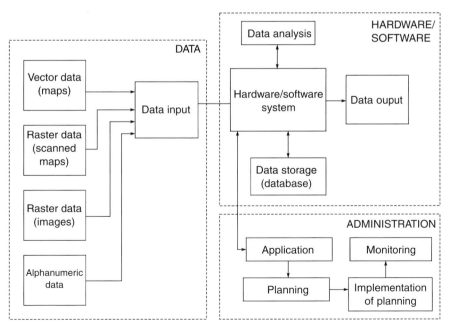

Figure 4.1 Concept of a geographic information system.

includes the data, which are managed by an administration or a unit conducting a project for the purposes of data inventory, data analysis and data presentation for administrative support or for decision support.

The information system is based on data which are available in various forms:

- spatial objects are represented by identifiers. They can relate to points, lines or areas administered in vector form. The identification and organization of these objects in coordinate and vector form is subdivided into feature or object classes. This includes their spatial or topological relations in two or three dimensions.
- data in raster form are also included. A pixel may be assigned an object code, or it may simply consist of grey levels of an image or a digital elevation model.
- the vector or the raster data are also linked to non-graphic information specifying place names and object numbers, which in databases may further be linked to a great variety of coded or alphanumerical attributes (e.g. owners of a parcel, inhabitants of a house, characteristics of a utility feature, statistical data for a defined area).

General and specialized GIS systems have been designed for a variety of purposes:

- for environmental management and conservation.
- for defence and intelligence purposes.
- for governmental administration.
- for resource management in agriculture and forestry.
- for geophysical exploration.
- for cadastral management.
- for telecommunications.
- for utility management.
- for business applications.
- for construction projects.

Many of these applications require common base data. It is the purpose of an administrative authority to create a spatial data infrastructure by which the base data may easily be exchanged. The main preoccupation in this context is the creation of a topographic database onto which thematic data of specific interest may be added (see Figure 4.2).

The justification lies in the fact that cost–benefit studies of thirty-nine GIS projects in Scandinavia, carried out by Nordisk Quantif, have shown that automation of a single production task in an administrative unit results in a cost–benefit ratio of 1:1, while the integration of data with other organizations can raise this ratio to 1:4.

This is all the more significant since, on average, the cost of data acqui-

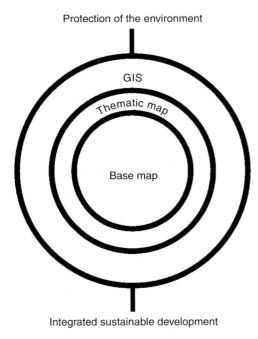

Protection of the environment

GIS

Thematic map

Base map

Integrated sustainable development

Figure 4.2 The relationship between base data and thematic data.

sition and the effort for its updating by far exceeds the costs for hardware, software and data processing.

Throughout the development of GIS systems, the hardware costs steadily went down. After a rise in the 1980s, the software cost has likewise taken a downward turn. With increased hardware and software power, GIS and data management likewise get more efficient and cheaper. What remains high in cost is the provision of data, particularly if these are to be kept up-to-date to reflect a model of the actual geographic and socio-economic environment.

The GIS pyramid of Figure 4.5 reflects the need for three types of users involved with GIS.

Most users are involved solely in the viewing aspects of GIS data. A smaller group is involved in the analysis of data. This is done by modelling to arrive at information deduced by combining different GIS data sets. A small group of GIS users is involved in base, thematic and attribute data provision and its updating.

Figure 4.6 shows the evolution of a GIS system over a number of years. Initially the focus is on providing an inventory of data. Later, analysis is highlighted. Finally, the emphasis is on management.

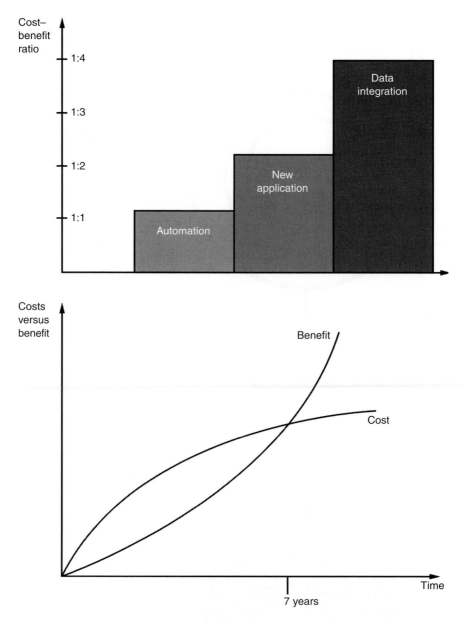

Figure 4.3 Cost–benefit ratio of GIS projects.

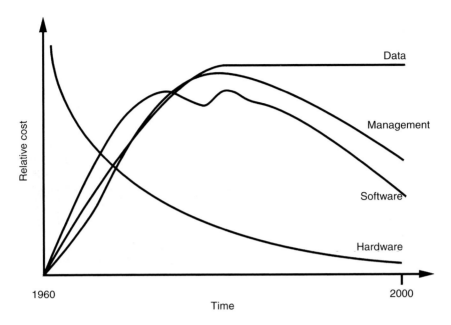

Figure 4.4 GIS cost aspects.

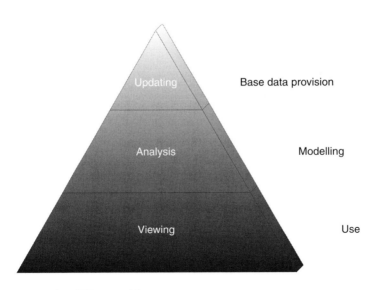

Figure 4.5 GIS pyramid.

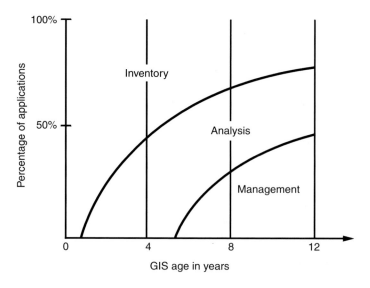

Figure 4.6 Evolution of a GIS.

Hardware components

The hardware of a GIS is composed of:

- input devices.
- processing and storage devices.
- output devices.

Input devices

Digital data input depends on the type of data to be utilized.

Imagery input is possible from analogue images through the use of image scanners. Digital airborne and space-borne systems already use charge-coupled device CCD-sensors to supply the data in digital form. Light falling onto a semi-conductor is transformed into an electric charge and into electric current. The light energy is proportional to the electric current, and thus brightness measurement becomes possible.

Area CCDs are capable of providing a full frame transfer of a shutter released image at full resolution. However, they suffer from long read-out times. Future CMOS technology may overcome the current size limitations of area CCDs. High-resolution systems therefore prefer the use of long linear arrays operated as push-broom scanners, which integrate charges and read them out line by line, without the use of a shutter.

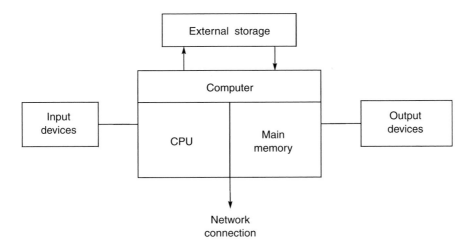

Figure 4.7 Hardware components.

For analogue and for digital images a resolution of 50 lp/mm can be reached.

Maps can be manually digitized by two-dimensional digitizers in vector form. This is possible in a single point mode or by dynamic measurement based on distance or time. The resolution of digitizing is about 0.2 mm. This is achieved by a fine wire grid inside the digitizing table. Digitizers are available for an A2 format or larger.

Maps may, however, also be raster scanned using scanners. These are available as drum scanners or as flat-bed scanners with a pixel size of 7 μm and an accuracy of 2 to 4 μm. They range from inexpensive desktop scanners limited in geometric and radiometric resolution to expensive but accurate cartographic scanners varying in geometric and radiometric resolution, their transparent or opaque use, and their suitability for black and white or for colour scanning. Scanned raster data may subsequently be converted into vector information with GIS software.

3D-vector data can be obtained directly by terrestrial survey equipment, such as:

- theodolites.
- electronic tacheometers.
- levelling instruments.
- GPS receivers.
- mobile mapping systems.

3D information from aerial photographs may be compiled by analogue or analytical plotters or by digital photogrammetric workstations.

The advantage of manual 2D or 3D *vector digitization* is that non-graphic attribute data may easily be attached via keyboard or menu. When manually digitizing vector information from images, the following limitations of the human eye should be considered. The monoscopic acuity of the human eye at a viewing distance of 25 cm is about 7 lp/mm. For an imaging system with a resolution of 60 lp/mm, this permits a viewing magnification of about 8 for a 7 μm pixel size of image. Current computer screens with 1664×1248 pixels only have a resolution of 300 μm per screen pixel.

Considering that the stereoscopic acuity of the eyes is even better, this means that screen digitization is inferior to 3D digitization (line extraction) on analytical plotters. Digital photogrammetric workstations, however, offer advantages for automated operations.

Processing and storage devices

Processing and storage devices consist of the central processing unit (CPU) and the main memory, the external storage devices and the user interface (see Figure 4.8).

The CPU executes the program commands. Its arithmetic unit performs algebraic and logical operations for the data. Its control unit regulates the data transfer between arithmetic unit and the main memory.

The *main memory* (random access memory, or RAM) contains the machine programs and accepts data in short access time with caching, if required. The *I/O controller* communicates with the periphery for hardware ports and for software drivers. The *bus system* establishes the con-

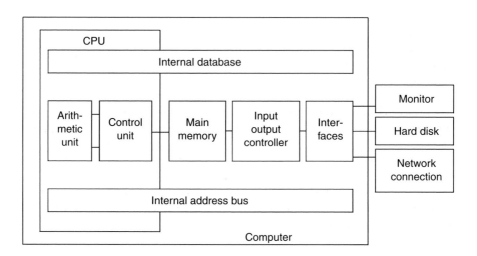

Figure 4.8 CPU and main computer memory.

nections. To speed up the output process, additional graphic cards and memory are usually added as interfaces.

Criteria for the CPU's performance are:

- the processor speed (over 300 MHz).
- the internal data format (32 or 64 bit).
- the external data format between the CPU and the main memory (64 bit).
- the physical memory (over 4 GB).
- the computing performance (over 2000 MIPS).

External storage devices are linked to the computer. The following options are available:

Diskettes of 1.44 MB.
ZIP tapes of 100 MB.
Magnetic hard disks of >30 GB with <10 msec access.
For archival magnetic tapes, CD-ROMs for 650 MB up to 3.2 GB are available with access speeds of <150 msec.
Jukeboxes are larger archival devices for a storage capacity of several terrabytes (TB).

For the current GIS systems, the following computer configurations can be considered:

PCs with one or several Intel Pentium Processors for use with Windows operating systems.
Workstations by
Hewlett-Packard: RISC
Sun: SPARC
IBM: RISC
DEC: α
SGI: MIPS

PCs or workstations are usually networked together in a client–server arrangement. The *server* holds the data in files or databases and contains application programs. The *client* is a user terminal (PC). The network is administered in a Local Area Network (LAN) or in a Wide Area Network (WAN).

The *user interface* consists of a high-resolution CRT screen, which may be adapted for colour viewing and optionally for stereo. It also consists of a mouse and a keyboard.

Output devices

Output devices include the ports to printers. Specific to GIS are the following graphic output facilities.

Vector devices are flat-bed plotters and drum plotters. Flat-bed plotters have an accuracy of ±0.05 mm at a speed of <30 m/min operated with a pen or a light beam. Drum plotters are less accurate but faster (300–900 m/min). They are used for verification plots.

Raster devices permit the output of halftones in a pixel or a screened manner. They are able to print RGB or CYMK colours in different saturations. They can combine vector and raster data in raster form. To print halftones, the dithering technique is used, in which printer pixels are combined in halftone cells. For example, a 600 dpi output has a 150 dpi halftone cell for a 4-bit radiometric resolution.

Screened colour reproduction uses different screen angles:

0° for yellow.
15° for magenta.
45° for black.
75° for cyan.

Laser raster drum plotters are available for film printing up to the A0 format with a resolution of 0.01 mm. Other output possibilities are by dye sublimation, thermal wax transfer and inkjet technology or laser printing.

A typical configuration of a GIS hardware installation is shown in Figure 4.9.

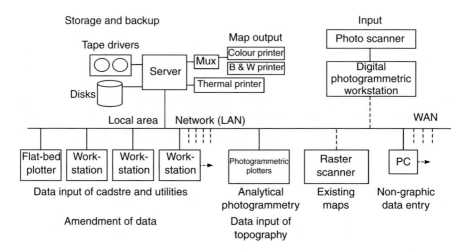

Figure 4.9 Typical GIS hardware installation.

Software components

Operating systems

The operation of a computer is based on its operating system. It assures that all parts of the computer function in liaison. Most common are Microsoft's operating systems for PCs.

In *MS-DOS* (Microsoft Disk Operating System) the operation is regulated by text lines. This permits the administering of files by name. More modern are *Windows* operating systems such as Windows 3.1, Windows 95, Windows 98, Windows NT, Windows 2000, Windows ME and Windows XP, utilizing graphic symbols (icons). Windows acts as a graphical user interface (GUI). Windows is now a network compatible system.

An operating system for workstations is UNIX, which has been adapted for the computers of specific manufacturers: HP-UX by Hewlett-Packard, AIX by IBM and Linux, which is generally available. Unix development dates back to the 1960s. It was originally designed for the operation of mainframe computers with multitasking. It contains a great number of data security features regulating access.

Programming languages

The programming of computers is made possible by programming languages, which translate user formulations into a machine-compatible code. For this translation a compiler for the respective programming language is required. Most GIS programs, based upon a chosen operating system, have been programmed in the programming language Fortran (formula translation). More modern languages are C, C++, and Visual Basic.

Networking software

The communication of computers within a local area network (LAN) and a wide area network (WAN) is assured by International Standards Organization (ISO) standards. The most common standard is TCP/IP (Transmission Control Protocol/Internet Protocol). TCP/IP separates data transmission into smaller packages transmitted from an identified sender to a receiving computer. The transmission of the packages is checked during the process.

Graphic standards

Graphic standards have been introduced so that the complex graphic instructions of the computer can be translated into monitor-compatible instructions. An internationally-agreed graphic standard is the Graphic Kernel System (GKS). It defines 2D graphic primitives (position, height, line type, font, colour and fill). Other standards are X window (X11) and special standards for 3D graphic cards.

GIS application software

Based upon an operating system, augmented by additional programming tools and standards, various vendors (ESRI, Intergraph, Siemens and many others) have developed GIS software packages. They have a great number of elements in common:

- translation (translation, rotation and scale change in the two dimensions of the screen).
- polygon creation (linking a line network to the origin of the line string).
- adjustment of polygons (observing conditions of right angles and parallel lines).
- line smoothing (connecting line strings by curves).
- vector to raster conversion (for display of vector information).
- raster to vector conversion (line derivation of vectors from pixels representing a line).
- edge cut-off (to fit a seamless data set to the screen).
- edge matching (to fit lines of adjacent tiles together).
- geometry edit (to change point and line information).
- intersections (between point locations, lines and polygons) in one layer.
- intersections between layers.
- buffer zone generation for points, lines and areas.
- counting of points in areas.
- measurement of point coordinates, distances and areas.
- interpolation.
- modelling functions.
- network analysis.
- symbol and text generation.
- generalization.
- map annotation.

These tasks will be discussed for vector and for raster systems (see Figure 4.10).

Vector systems

Object representation

The modelling of vector geometry depends on local or georeferenced coordinate systems. The advantage of vector systems lies in the possibility of

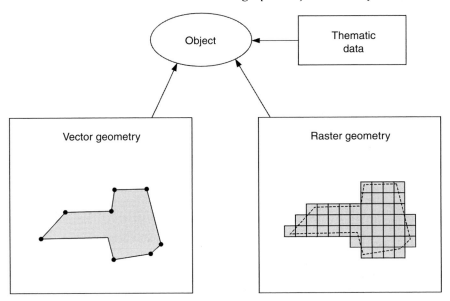

Figure 4.10 Vector and raster geometry.

recording and displaying coordinates with full measurement accuracy of ground surveys or of photogrammetric point and line measurements. In general, vector systems also contain less data volume than raster images of the same area. Furthermore, it is easy to attach alphanumeric attributes to the defined elements of a vector system (see Figure 4.11), such as:

- points.
- lines.
- areas.
- objects.

A point is defined by its coordinates, x, y, and by its node number. A line is defined by the coordinates of its end points, x_1, y_1 and x_2y_2 and its line (arc) number. A line string is defined by the coordinates of all points forming the line string: $x_1y_1, x_2y_2, \ldots x_ny_n$. An area is defined by the coordinates of the line string ending at the initial points: $x_1y_1, x_2y_2, \ldots x_{n-1}y_{n-1}, x_1y_1$. To points, lines and areas, attributes with alphanumeric thematic data may be attached (see Figure 4.12).

In Figure 4.13 the different ways in which an area may be represented are shown.

B shows the digitization in the computer-aided design (CAD) form of 'spaghetti graphics'. CAD systems such as Autocad or Microstation in their simplest form do not automatically snap adjacent lines to common

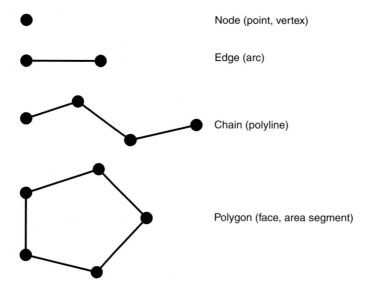

Figure 4.11 Point, line and area objects.

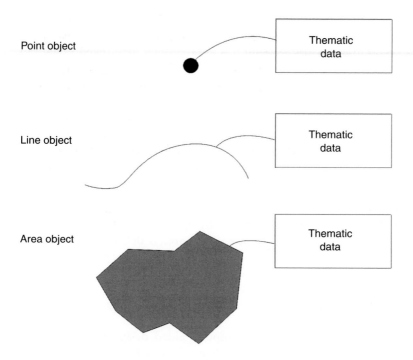

Figure 4.12 Attribute links.

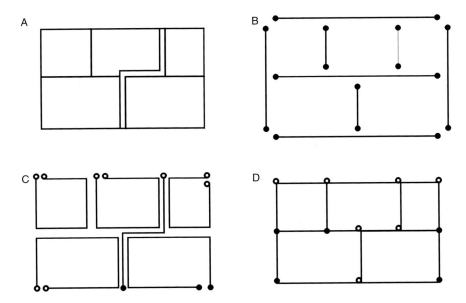

Figure 4.13 Graphic representation of an area.

end and intersection points. Their result is a visual area representation, which cannot be analysed for adjacency.

C shows the representation of areas by formation of closed individual polygons and by selection of additional line strings. Intergraph's MGE software provides this partial topology model.

D shows the generalized attempt to digitize areas through the formation of polygons and by intersecting them with lines. In this way, a relational geometry model may be built up if the rules of topology are observed.

The relations are expressed in three relational tables. These relate to Figure 4.14.

A line-area table

Line	Area left	Area right
a	⓪	①
g	②	①
e	⓪	①
f	⓪	①
b	⓪	②
c	⓪	②
d	⓪	②
g	①	②

A coordinate table for all points

Point	Coordinates
1	x_1y_1
2	x_2y_2
3	x_3y_3
4	x_4y_4
5	x_5y_5
6	x_6y_6

A line-point table

Line	From point	To point
a	1	2
b	2	3
c	3	4
d	4	5
e	5	6
f	6	1
g	2	5

Please note that the outside area is also identified, and that the direction of the line must be indicated to assure full topology. The relational model may be contained in a large relational database (Oracle Spatial, Siemens) or it may be administered separately (ArcInfo by ESRI).

The knowledge of the topological relations permits neighbourhood queries. Non-graphic attributes may be directly attached to the identifiers of points, lines and areas or to objects composed of areas or line strings.

Early GIS developments favoured the CAD or partial topology models, because they needed less computing power and could operate on larger

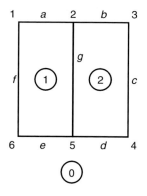

Figure 4.14 Topological model.

databases. Attributes were then attached to 'pointers', which consisted of points identifying the graphic placement of an area (parcel) number, constituting the link to the non-graphic database content. In partial topology systems, the area number could directly establish the link to the attributes.

Since it was easier to digitize in CAD systems, topology was created by software snapping to common points and to line intersections in a post-digitizing batch process assuming a tolerance distance. In doing so, point numbers, line numbers and area numbers were automatically created and the relational tables were set up accordingly.

GIS systems adopted the *layer concept*, in which graphic information of a certain theme was stored in separate graphic files. These were easier to manipulate. Different layers could be superimposed in separate colours, much like the themes in a topographic map printed in different colours. The intersection of two different layers presented additional topology generation needs through a subsequent batch process. The generation of a GIS in an object-oriented manner is helpful in this. The German Topographic System ATKIS groups the elements of topographic objects into hierarchical categories:

code 1000   control points
code 2000   buildings
code 3000   transportation
code 4000   vegetation
code 5000   hydrography
code 6000   topographic relief
code 7000   boundaries

Each category is divided into object groups, for example:

code 3100   roads
code 3200   railways
code 3300   air traffic
code 3400   ship traffic
code 3500   buildings for traffic purposes

Then each object group is divided into types, for example:

code 3101   road
code 3102   path
code 3103   square

Each object type, group or category may contain its specific non-graphic attributes (see Figure 4.15).

If the geometry of a defined object extends over several layers, we speak

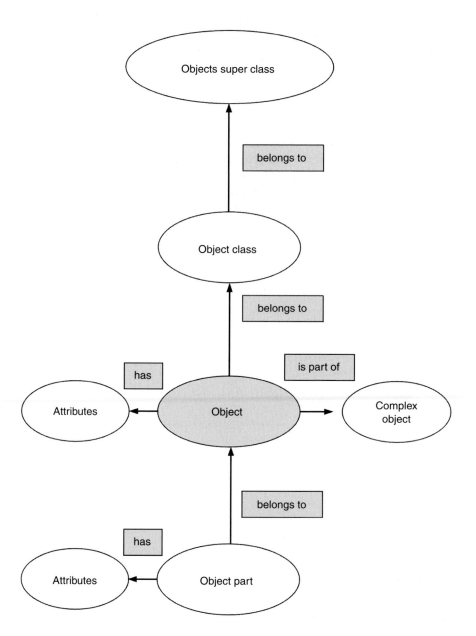

Figure 4.15 Object orientation.

of an 'object oriented GIS'. Attributes may be inherited for these objects, and specific processing methods may be applied to them.

In the LH-Systems/Laser Scan Lamps 2 software, parts of the database in different layers may be commissioned by object orientation to different users.

Vector geometry

The calculation of vector positions and intersections follows the principles of analytic geometry. Depending on the two- or three-dimensional representation chosen for a GIS, two- or three-dimensional analytic geometry is utilized for the formulations. In two dimensions, the following terms may be calculated from coordinates:

Distances:

$$d_{ij} = \sqrt{(x_i - x_j)^2 + (y_i - y_j)^2}$$

Directions (for a system in which x points up and y to the right):

$$\alpha_{ij} = arctg \frac{y_i - y_j}{x_i - x_j}$$

New point positions:

$$x_j = x_i + d_{ij} \sin \alpha_{ij}$$

$$y_j = y_i + d_{ij} \cos \alpha_{ij}$$

Angles:

$$\beta_{ijk} = \alpha_{ik} - \alpha_{ij}$$

Areas of a polygon:

$$A = \frac{1}{2} \sum_{i=1}^{n} (x_i + x_{i+1})(y_{i+1} - y_i)$$

Centre of mass of a polygon:

$$x_c = \frac{1}{6A} \sum_{i=1}^{n} (x_i + x_{i+1})(x_i y_{i+1} - y_i x_{i+1})$$

$$y_c = \frac{1}{6A} \sum_{i=1}^{n} (y_i + y_{i+1})(x_i y_{i+1} - y_i x_{i+1})$$

Circumference of a polygon:

$$C = \sum_{i=1}^{n} \sqrt{(x_i - x_{i+1})^2 (y_i - y_{i+1})^2}$$

For three dimensions, these expressions are expandable as stated for solid analytical geometry.

Geometric spatial queries such as those shown in Figure 4.16 depend on calculations of intersections between points, lines and areas.

The equation of a straight line in two dimensions is given by

$$Ax_i + By_i + C = 0$$

Using the coordinates of the two endpoints defining the line, the formulation is:

$$\frac{y_i - y_1}{y_2 - y_1} = \frac{x_i - x_1}{x_2 - x_1}$$

or

$$\begin{vmatrix} x & y & 1 \\ x_1 & y_1 & 1 \\ x_2 & y_2 & 1 \end{vmatrix} = 0$$

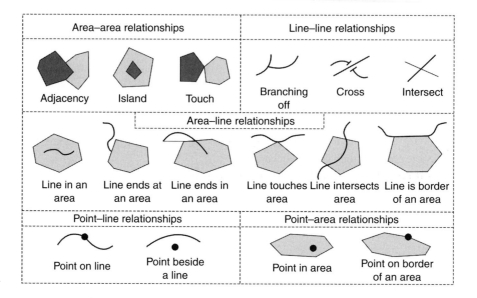

Figure 4.16 Geometric spatial queries.

The shortest distance, d, of a point, P, from a straight line becomes:

$$d = \frac{Ax_p + By_p + C}{\sqrt{A^2 + B^2}}$$

Two lines

$$A_1x_i + B_1y_i + C_1 = 0$$

$$A_2x_i + B_2y_i + C_2 = 0$$

may be intersected. The point of intersection has the coordinates:

$$x_i = \frac{B_1C_2 - B_2C_1}{A_1B_2 - A_2B_1}$$

$$y_i = \frac{C_1A_2 - C_2A_1}{A_1B_2 - A_2B_1}$$

The straight lines are parallel if $A_1B_2 - A_2B_1 = 0$.
They are perpendicular if $A_1A_2 + B_1B_2 = 0$.
The distance, d, between the parallel straight lines is:

$$d = \frac{C_1 - C_2}{\sqrt{A_1^2 + B_1^2}}$$

The equation of a circle is:

$$x_i^2 + y_i^2 + Ax_i + By_i + C = 0$$

or

$$(x_i - x_o)^2 + (y_i - y_o)^2 = r^2$$

Using the tools of analytic geometry, these formulations may be expanded to second and higher order curves.

In three dimensions, the equivalent terms become:
Distances:

$$d_{ij} = \sqrt{(x_i - x_j)^2 + (y_i - y_j)^2 + (z_i - z_j)^2}$$

The direction cosines:

$$\cos \alpha_x = \frac{(x_i - x_j)}{d_{ij}}$$

$$\cos \alpha_y = \frac{(y_i - y_j)}{d_{ij}}$$

$$\cos \alpha_z = \frac{(z_i - z_j)}{d_{ij}}$$

The spatial angle between directions with the direction cosines α_x, α_y, α_z and α_x', α_y', α_z' is:

$$\cos \beta = \cos \alpha_x \cos \alpha_x' + \cos \alpha_y \cos \alpha_y' + \cos \alpha_z \cos \alpha_z'$$

The equation of a plane determined by three points becomes:

$$Ax_i + By_i + Cz_1 + D = 0$$

or

$$\begin{vmatrix} x_i & y_i & z_i & 1 \\ x_1 & y_1 & z_1 & 1 \\ x_2 & y_2 & z_2 & 1 \\ x_3 & y_3 & z_3 & 1 \end{vmatrix} = 0$$

A straight line in three dimensions is defined by two independent linear equations:

$$A_1 x_i + B_1 y_i + C_1 z_1 + D_1 = 0$$

$$A_2 x_i + B_2 y_i + C_2 z_1 + D_2 = 0$$

or

$$\frac{x_i - x_1}{x_2 - x_1} = \frac{y_i - y_1}{y_2 - y_1} = \frac{z_i - z_1}{z_2 - z_1}$$

A distance, d, between a point, P, and a plane is given by:

$$d = \frac{Ax_p + By_p + C_p + D}{\sqrt{A^2 + B^2 + C^2}}$$

So far, most GIS systems have been limited to two-dimensional geometry. Elevations have, however, been included as attributes. One therefore speaks of a two-and-a-half-dimensional capability. The elevation information may be introduced on a case-by-case basis, if a special three-dimensional query is desired and programmed.

With the described tools of analytic geometry, the GIS vector application software tasks listed in this chapter (page 194) may be programmed, for example:

Translations, Δx and Δy, are possible by:

$$\begin{pmatrix} x' \\ y' \end{pmatrix} = \begin{pmatrix} x + \Delta x \\ y + \Delta y \end{pmatrix}$$

Rotations by an angle, α, are determined by:

$$\begin{pmatrix} x' \\ y' \end{pmatrix} = \begin{pmatrix} \cos\alpha & -\sin\alpha \\ \sin\alpha & \cos\alpha \end{pmatrix} \begin{pmatrix} x \\ y \end{pmatrix}$$

Scale change by a scale factor, λ, is executed by:

$$\begin{pmatrix} x' \\ y' \end{pmatrix} = \lambda \begin{pmatrix} x \\ y \end{pmatrix}$$

Edge cut-off is calculated by the calculation of intermediate points intersecting the equation of a line with the boundaries of a rectangle, $x_{max}y_{max}x_{min}y_{min}$, of the area to be displayed.

Raster systems

Raster data consist of a regular 2D grid of square cells. The grid is characterized by a (geocoded) origin, its (geocoded) orientation and the raster cell size, which for imagery corresponds to a pixel (picture element) size. Other information, such as elevation levels or thematic data, may also be arranged by a scheme of regular tessellation. Raster systems may also be arranged in three dimensions. The 3D cell becomes a cube (a voxel). The attribute of the cell describes the thematic information (grey level, elevation level, thematic object content). Raster coverages may be in regular (square, rectangular) or irregular dimensions. Each raster data set constitutes a layer. There may be many layers for the same area.

The geometric accuracy of raster data is limited by the cell resolution. A mixed-cell problem may exist. Due to limits in resolution, there is a possibility of mixed pixels.

While raster models reflect what is present, vector models more accurately define the whereabouts. Raster topology is defined by the eight neighbouring pixels surrounding a particular pixel. Neighbouring cells carrying the same attribute define a connection component. In this way linear objects may be recognized by the connection components.

Raster data operations are possible in the following ways:

- *geometric transformations*, which permit geocoding via digital (differential) rectification and the presentation of perspective views including its geometric resampling algorithms.
- *radiometric transformations*, which include all types of digital image processing (filtering, multispectral classification, image analysis).

Geometric and radiometric transformations have already been discussed in previous chapters.

- *algebraic transformations*, which permit the combination of different layers and their analysis by Boolean operators. Boolean operators are: AND, OR, XOR and NOT. AND signifies if both layer pixels have a particular value. OR signifies if one of them has that particular value. XOR means the exclusive OR, that both do not have that particular value, and NOT signifies that one of them does not have that particular value.
- *macro-operations*, applied for line enhancement or line thinning, are the Blow and Shrink operations. They consist of a shift of the raster image in all four directions and the logical OR application. This thickens the line. For line thinning the same operation is applied for the background. Holes which may have resulted from the operation are filled by filtering.

Databases

A database is a self-contained, long-term organization of data for flexible and secure use. It consists of the data and of a database management system, the software to manage the data. A database permits a strict separation between data and an application. It has a well-defined interface for application programs. The user of a database is not concerned with the internal data organization, but he or she can change the data location without changing the application program. The database management system provides efficient access to the data with security checks.

The internal view of a database is the physical memory allocation. The conceptual view concerns the logical data organization and the external view to the user is the graphic user interface. Databases are used in a large variety of applications (banking, reservation systems, libraries, business). GIS can take advantage of these developments.

Basic to the generation of a database is the entity–relationship model for a particular application. The relationship defines the association between entity types, for example:

point number – coordinates   1:2
building – parcel n:1
parcel – village m:1

Meta Data are stored together with the data. They, for example, define the reference system, the resolution and the date of data acquisition.

Database structures may be

- hierarchical.
- network type.
- relational.
- object–relational.

Hierarchical databases are the oldest database type. They support 1:*n* relationships well, but lead to redundant storage for *n:m* relationships. They are based on a tree structure, and they are inflexible with regard to changes.

Network databases have been available since the 1970s, when the Conference on Data System Languages (CODASYL) introduced the first commercially available database of this type. It supports 1:*m* and *n:m* relationships without redundancy. It was used in older GIS systems.

The relational database is currently the most common database model. It is based on relational tables. Each entity is characterized by a table, called a 'tuple'. Relationships between different entity types are also expressed by tables. The sequence of tuples is irrelevant. Unique access to an individual entity is made possible by a segment having a 1:1 or *n*:1 relation to other segments (e.g. point number, position). The organization of each table is independent of other tables. Tables can be accessed, combined and changed by simple operations. Masking permits different external views of the data. Simple rules prevent redundancy. The disadvantage of relational databases is that they are designed for simple data. For 2D topological and for 3D spatial queries they become slow for interactive work. Thus at least temporary extensions in GIS systems can accelerate the management of interactive sessions. Microsoft's 'Data Access' is a typical relational database.

Object–relational databases are extensions of relational databases towards object orientation. They contain user-defined data types as domains of a table and user-defined functions. Examples for object–relational databases are:

- Oracle 8   Spatial Data Cartridge.
- ESRI   Spatial Database Engine SDE.
- Informix   2D-Spatial Data blade.

Many public data collections have a spatial reference in the form of a location name, called 'indirect spatial access', e.g. street address, city.

GIS data have a 'direct spatial reference' via 2D or 3D coordinates. To use indirect spatial access, data in GIS geocoded links for these must first be established (e.g. by a parcel number, a building number, or a set of coordinates). In very large databases, a coordinate reference is usually added to the data to allow for efficient retrieval.

The geometric and thematic data of a certain area are stored together in regular raster cell divisions of the database. This map-based (tile-based) organization assures quick access to the data (Arc-Info).

Another possibility is to organize the raster cells in the form of a quad tree structure. The objective is to store approximately equal amounts of data content in each cell. If a regular raster cell in a coarse raster cell division contains more data than a limit allows, then the raster cell is subdivided consecutively in half or quarter dimension cells. Spatial indexing is used for quick access of the data sets (Sicad, Oracle Spatial). Other cell subdivisions, such as the *k*-dimensional tree (K-D tree) or the grid file method with irregular raster cell dimensions are possible. The R-Tree (Intergraph-Tigris) utilizes a minimum-bounding rectangle for entities and groups of entities determined by clustering.

GIS systems

There are a few hundred GIS systems in existence. The following website contains useful information on these systems:

www.geo.ed.ac.uk/home/giswww.html.

Websites of some of the major international GIS vendors are:

Bentley, USA	www.bentley.com
Caris, Canada	www.caris.com
ER Mapper, USA	www.ermapper.com
Erdas, USA	www.erdas.com
ESRI, USA	www.esri.com
Genasys, Australia	www.genasys.com
GE-Smallworld, UK	www.smallworld-us.com
Idrisi, USA	www.idrisi.clarku.edu
Intergraph, USA	www.intergraph.com
MapInfo, USA	www.mapinfo.com
PCI Geomatics, Canada	www.pci.on.ca
Sicad, Germany	www.sicad.de

On these websites, the various software components offered by these companies can be traced.

The early development of commercial systems was conducted separately for vector and raster systems. ESRI produced Arc/Info, with a graphic database Arc permitting a topological structure and a non-graphic database Info. The major contender, Intergraph, produced a VAX-based system with a CAD structured graphic database IGDS and a non-graphic database DMRS. Due to the rapid development of computer technology towards server–client systems, ESRI has augmented Arc/Info with the additional database SDE. At the same time, Arc/View was developed for desktop computers, with access to raster data. At Intergraph, developments made use of Bentley's Microstation CAD programs to be utilized in the MGE environment. Special capabilities for spatial analysis with a topological structure were added with MGA. The integration of raster data with the vector data was also made possible with Imager (I/RAS). For the treatment of large volumes of data, MGDM was developed, which could split large area coverages into domains consisting of themes. The product TIGRIS developed for large military databases found its way into the civilian world under the name of DYNAMO in a topologically structured database environment. Finally, the graphic CAD base of Microstation was replaced by Imagineer for the Geomedia products of Intergraph.

The originator of Microstation, Bentley has in the meantime produced Geographics as a complete Bentley GIS product. SICAD of Germany has had long-standing GIS developments in conjunction with the German governmental topographic and cadastral databases ATKIS and ALK, now extended into ALKIS. Among the smaller ESRI and Intergraph competitors, Smallworld has been successful. Rather than separating the GIS into a graphic and a non-graphic database part, Smallworld stores graphic and non-graphic data together. Caris is another successful contender, particularly for hydrographic, cadastral and geological applications. The development of raster systems was pushed forward by Erdas Inc., by Idrisi and by PCI Geomatics.

In the meantime, all major GIS system vendors have either their own possibilities or interfaces for the integration of raster and vector data. A prime example is the cooperation between ESRI and Erdas. PCI Geomatics offers interfaces between their raster products and the Canadian vector system producers, Spans and Pamap, but also to Ilwis generated by the ITC Netherlands. For the integration of photogrammetric data systems of Erdas, LH Systems, PCI Geomatics and Z/Imaging, strong cooperations have been developed between ESRI and ERDAS, but also between ESRI and LH-Systems. Intergraph has strong ties with Z/Imaging.

The GIS vendors have realized the weakness of producing their data in proprietary formats. Raster systems TIFF (Tagged Image File Format) and GeoTIFF format (with metadata for georeferencing) presented few difficulties. These formats may be compressed into the JPEG (Joint Photographic Experts Group) format at variable compression rates.

The vector data format conversion, for example, between ESRI formats (dxf files), Intergraph formats (dgn files), SICAD formats (EDBS files) or AutoCAD formats (dwg files) necessitated more or less effective conversion programs. The Open-GIS Consortium, in which the vendors of GIS systems cooperate in attempts to assure transferability of data generated by the particular vendor system into that of another vendor. Various working groups of the International Standards Organization (ISO), in particular of the Technical Committee TC 211 on Geographic Information and Geomatics, support this effort.

GIS and the Internet

Most GIS vendors have already provided access to an Intranet or the Internet, for example:

- Internet Map Server and Map Objects by ESRI.
- Geomedia Web Map by Intergraph.
- Smallworld Web by Smallworld.

The Intranet and Internet do not differ in technology but they regulate access. The Intranet is limited to use by a company or an administration, while the Internet is accessible, for example, through the worldwide web, often at a cost.

The data reside in Geodata servers and are remotely accessed. The client in the web accesses the server via plug-ins or applets. These are small programs, which can be loaded from the Internet. Java applets are independent of a specific platform. The server contains a common gateway interface. The client makes a request at the web or ftp-server through the web browser. The request is passed on to the common gateway interface to start the application (e.g. to retrieve a particular data set). The answer is passed on through the gateway interface and the web or the ftp-server to the client. Ftp-servers permit the exchange of large data sets. The connection between client and server is through wired telephone connections which, depending on the particular service, can vary in performance from 19 Kbaud (19 kb/sec) to 64 Kbaud for ISDN connections, to several Mbaud for Ethernet or glass fibre connections.

In the future, wireless connections will be offered by mobile telephone companies. This will reach a full impact when, for example, the European GSM system is replaced by the UMTS mobile phone system, permitting higher transmission rates. The transmission of data for wireless connections is regulated through the wireless application protocol (WAP). The mobile phone industry and the industry producing personal digital assistants (PDA), such as Hewlett-Packard, Nokia, Palm and Psion, have provided possible opportunities for mobile GIS users.

GIS applications

GIS applications are scale dependent.

Global applications

The topographic base data for global applications is the digital chart of the world (DCW) compiled by the US National Imaging and Mapping Authority (NIMA) and distributed at low cost through the US Geological Survey (USGS) at the scale 1:1 000 000. It has been integrated from available national data sets. DCW data can be purchased on CD-ROM. They are also downloadable from an ftp-server with special software to be purchased via ftp://216.15.110.75/downloads/DCW/DCWdata/.

Digital elevation models, also compiled by NIMA, are downloadable at 1×1 km spacings through NOAA: www.ngdc.noaa.gov/seg/topo/globe.shtml.

Based on these references, a variety of satellite image data are rectified, monitoring the atmosphere, the land and the oceans. Typical global datasets are the NDVI for the monitoring of vegetation (see: http://earthobservatory.nasa.gov).

On a continental level, ESRI, in cooperation with the World Resources Institutes (WRI), has provided the African Data Sampler, in which various small-scale thematic coverages can be compared. Satellite images and thematic vector information can be merged for information at the global level in the form of a digital atlas (see www.earth-info.org).

To permit global access to geographic data for the actions of international organizations and continental activities, a global spatial infrastructure (GSDI) is required. For its implementation, a GSDI international committee structure has been created. According to an 'SDI-Cookbook' available through www.gsdi.org, the exchange of data through world base clearing house nodes is in preparation. Initiatives such as the 'Global Map' (1:1 000 000) and 'Digital Earth', propagating the technology, have been started. They are supported mainly by the USA, Canada, Australia, Japan, with European countries, China and others participating. The link to the former United Nations Cartographic Conferences has permitted the establishment of an Asia-Pacific SDI committee and a Latin American SDI committee.

National applications and spatial data infrastructure

The national supply of base data depends on the activities of the national mapping agencies at medium scales (1:25 000 to 1:50 000), which permit a more detailed GIS analysis. Since these agencies only provide the topographic base data, which must be supplemented and integrated with thematic data sets from other organizations, the lead agency for the exchange of data must strive for the establishment of a national spatial data infrastructure framework (NSDI). In the USA, a Federal Geographic Data Committee (FGDC) was formed in 1990, and an NSDI was introduced in 1994 by a Presidential Executive Order (no. 12906).

An NSDI not only contains the geodata in various forms (vector data of digital maps, raster data as digital orthophotos and as digital elevation models), but also the metadata describing the data sets. The clearing house provides access to them and constitutes a catalogue of the data. Parts of the NSDI are also the agreed framework agreements for the provision and updating of the different datasets according to agreed upon standards.

In most European countries, or in Canada, such national spatial infra-structures have been established by the national mapping agencies (e.g. Ordnance Survey in the UK, IGN France, Geomatics Canada) in liaison with user ministries. In countries where the large-scale mapping responsibilities have been assigned to the states of the country, the state survey administrations have created an NSDI by coordination committees (e.g. in Germany by the ADV). In Germany the national topographic information system ATKIS is completed for the entire country. Its geometry is derived from the digitization of the 1:5000 base maps (DGK) with an accuracy of ±3 m on the ground. So far, the system contains ninety object classes. It is possible to derive map data at a scale of 1:10 000 from the data sets covering the country.

Plate 19 in the colour section shows the GIS-derived topographic map, 1:25 000, extracted from ATKIS data.

Most European countries have countrywide orthophoto coverages at the 1 m ground pixel level. Some even have 0.5 m ground pixel coverage. Only Germany, with the exception of some states (e.g. Northrhine-Westfalia) does not have such coverage due to the existence of Atkis.

The base data provided and maintained by the national (or state) mapping agencies are being augmented by national, state or local authorities with their relevant thematic data for their uses for:

- regional planning.
- environmental management.
- statistical purposes.
- soil surveys.

Since the vector base data are more costly and less easy to manipulate than raster data, the mapping agencies have also supplied their raster scanned maps at scales of 1:50 000, 1:25 000, 1:10 000 and 1:5000. These data sets are easily available to local authorities and private companies, and they can create a value-added digital map for their specific purposes. They retain the responsibility for maintenance and updates of their data sets according to the agreed exchange standards.

A great number of value added applications are developable on a project basis such as:

- hydrological modelling (using vector data, raster images and DEMs).
- car navigation (with the need to provide monthly updates for traffic routes).
- business analysis (based on income of residents).
- accident and crime location (of interest to police and traffic planners).
- real estate valuation and real estate market.
- emergency planning (fire-fighting access, flooding).
- tourist information systems (location of stores, restaurants, public

places of interest, mailboxes, pharmacies, petrol stations, public phone booths, hospitals).

- health studies (occurrence of diseases in certain locations).

Local applications

Local decisions most often require large-scale information at map scales ranging from 1:100 for construction to 1:500 for utility location to 1:1000 for urban property cadastres to 1:2000 for rural property cadastres.

Cadastral applications

The existence and the maintenance of a cadastral system has historically proved to be advantageous. Introduced through Napoleon and his officers in the early nineteenth century, a cadastre served as a register of property for the purposes of property tax collection. It consisted of a large-scale map indicating the geometric location of land parcels with their parcel numbers. In the accompanying register, the property owners or users were recorded with a cross-reference to the parcel numbers of each district, village and block. The capital market, which permitted the owners of land to obtain loans on the basis of mortgages, made the cadastre a valuable asset to record these, together with land use rights and encumbrances in the non-graphic part of the cadastre (the 'book'). The graphic part of the cadastre (the 'map') permitted the administration of the neighbourhood relations of parcels, giving the owner the guarantee that the land parcel existed within the indicated dimensions without overlaps. Even though property boundaries were also monumented, the cadastral map proved useful in retracement surveys of these boundaries, in case monuments were lost.

In the early twentieth century, these property boundary points were tied into geodetic reference systems. The field survey records permitted the recalculation of the exact coordinates of the boundary points for the re-establishment of lost boundary markers. This permitted the establishment of a numerical cadastre of the parcel geometry in some densely populated regions, as opposed to a purely graphical cadastre, in which the boundaries were only recorded on the large-scale map.

The historical development of cadastral systems was different in the various regions of the world. While in Central Europe a numerical or at least a graphical cadastre was established, the English-speaking world favoured land title systems in which the geometrical references were contained in deeds or title documents.

With the advent of GIS, both types of property description and registration could be converted into a vector-based digital form. The cadastral GIS now contains graphical data in georeferenced form, describing the location and the dimensions of parcels and non-graphic attributes containing references to owners, land use and tax value. Emphasis is on a unique identifier, the parcel number, which is quickly locatable by the coordinate references. A cadastral GIS can now serve the purposes of:

- a tax cadastre.
- an ownership protection cadastre with the function of a land titles system.
- a multipurpose cadastre for the purpose of planning the local environment.

In Germany, the digital cadastral map ALK has been established on a 1:1000 or 1:2000 scale map base, uniquely identifying the parcel. The boundary points are georeferenced in point coordinate registers. A separate database exists for parcel numbers with the attributes (ownership, etc.) in the form of ALB. It is possible to obtain all relevant parcel records at the cadastral offices within minutes. By the year 2007, the ALK and the ALB are intended to be merged into a new cadastral GIS, with the name of 'ALKIS', for the entire country.

Figure 4.17 shows a part of the ALK map, 1:1000, depicting land parcels and boundaries, boundary points, buildings, land use, annotated land use and a limited number of topographic objects.

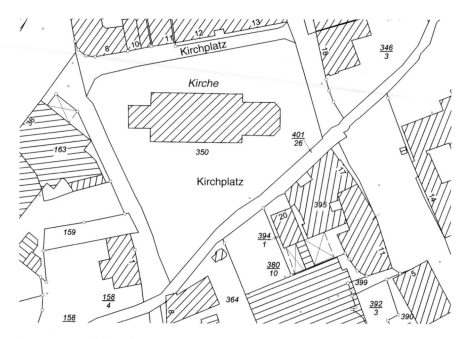

Figure 4.17 Cadastral map 1:1000.

Source: Map courtesy LGN, Hannover.

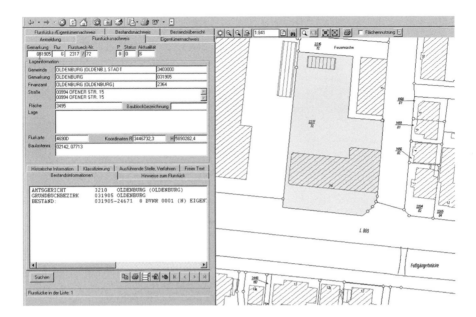

Figure 4.18 Ownership certificate.

Source: Map Hannoversch-Münden (ALK), courtesy of VKV Niedersachsen, Dezernat 207, Bez.-Reg. Weser-Ems, Oldenburg.

Figure 4.18 shows an ALKIS ownership certificate.

Figures 4.19 to 4.21 show value added applications of the cadastral GIS data for urban mapping.

The procedure for generating value added information is also shown in Figures 4.22, 4.23 and 4.24, in an example to derive access information for fire engines.

Recognizing the value of such a system for the economic stability of a country, the World Bank has supported many land management projects worldwide, which all contain land information systems on the basis of some form of digital cadastre. Typical examples are the Land Titling projects in Thailand with Australian consultancy, in Latin America with Canadian consultancy, in the reform countries of Eastern Europe with American, Scandinavian and German consultancies. These concentrate on low-cost approaches. If property or land use records do not exist, the location and the dimensions of parcels are established by photo adjudication, in which neighbouring landowners or land users agree on boundary locations identified in photographs or in orthophotos.

Plate 20 in the colour section shows the superposition of digital orthophoto with the relatively old cadastral map in Croatia with the need for the redetermination of boundaries.

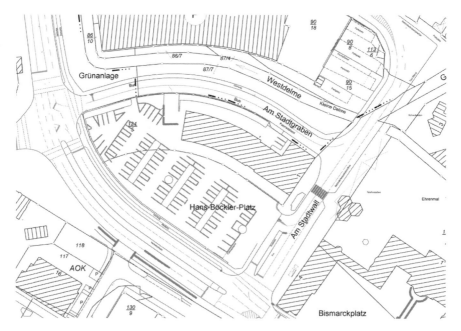

Figure 4.19 Cadastral map with road topography.

Source: Map Delmenhorst, courtesy of VKV Niedersachsen, Dezernat 207, Bez.-Reg. Weser-Ems, Oldenburg.

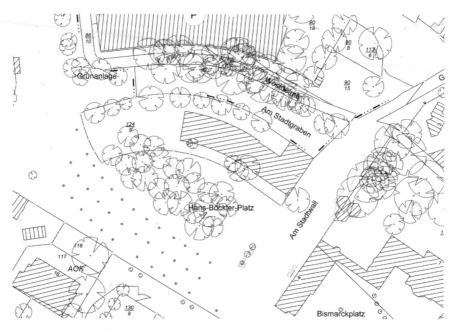

Figure 4.20 Cadastral map with urban vegetation.

Source: Map Delmenhorst, courtesy of VKV Niedersachsen, Dezernat 207, Bez.-Reg. Weser-Ems, Oldenburg.

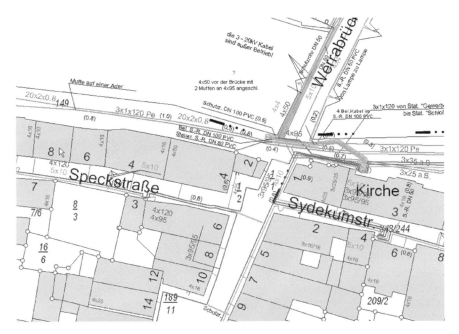

Figure 4.21 Cadastral map with utilities.

Source: Map Delmenhorst, courtesy of VKV Niedersachsen, Dezernat 207, Bez.-Reg. Weser-Ems, Oldenburg.

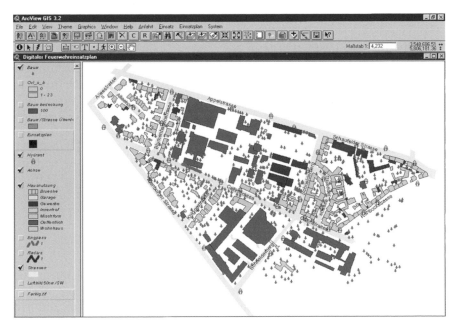

Figure 4.22 Cadastral GIS with minimal topographic information.

Source: Map courtesy of Institute for Photogrammetry and GeoInformation, University of Hannover.

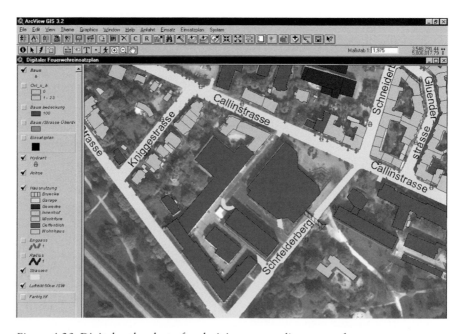

Figure 4.23 Digital orthophoto for deriving crown diameters of trees.

Source: Image courtesy of Institute for Photogrammetry and GeoInformation, University of Hannover.

In Africa, where in tribal land use rights prevail in many areas, it has been suggested to record these by pointers established on an orthophoto base, which link the existing rights in databases without the need to specify exact boundaries. This constitutes an effort to ensure sustainable development on the basis of information can take place within one generation.

Contrary to topographic GIS databases, which have a periodic updating need at intervals of a few years, cadastral GIS databases must be updated in near-real time. This imposes strict administrative procedures and efforts on behalf of governments assuring a secure land management for public and private uses.

Facility management

Another special GIS application area is in the inventory, management and planning of the utility infrastructure, concerning the distribution of electricity, telephone and computer line networks, the supply of fresh and irrigation water, of natural gas and the management of sewerage and rain water drainage systems.

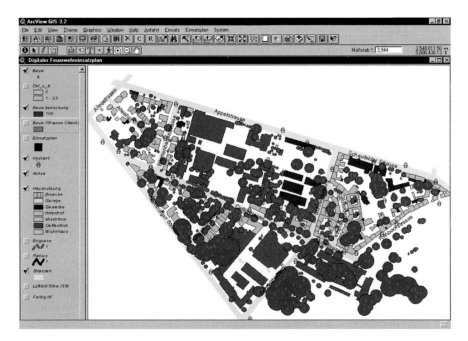

Figure 4.24 GIS with crown diameters of trees.

Source: Map courtesy of Institute for Photogrammetry and GeoInformation, University of Hannover.

A GIS permits the location of the manholes and the specific devices (distributor boxes, transformers, valves, etc.) geometrically. These are interconnected by cables or pipes in a linear network. Each utility network can be contained in separate graphic layers, which may be superimposed in a GIS, describing their relative location. Plate 21 in the colour section shows an urban utility network.

Often the utility administrators or companies prefer a schematic display of individual cables indicating their switching capabilities and their capacities. Attributes may describe depth of cables and pipes or their relative slope. Cable and pipelines may be linked into object-oriented networks, permitting the analysis of flows. Even the consumption of power, gas or water may be queried in these networks up to the point of billing.

City models

A relatively new requirement for the use of three-dimensional information from a GIS has developed from the mobile telephone industry. To locate the optical distribution of cellular phone antennas in urban areas they need three-dimensional city models to circumvent transmission obstructions caused by buildings. A standard base data set obtained by analogue, analytical or digital photogrammetry contains the 2D dimensions of the building footprints at the ground level of a topographic DEM. Additional manual photogrammetric measurements of the rooftops, which may also be obtained in an automated way by image matching techniques, permit the generation of 3D city models at various stages:

- the company Phoenics of Hannover has generated about 100 city models of the major German cities in a block form. Building polygons have been extracted from Atkis data. The building heights have been added via image correlation. This permitted the creation of the 3D building model as an input to the mobile phone antenna location programs.
- these models can be improved by the additional measurement of several roof types to permit schematic oblique views of the city.
- it is possible to integrate texture images taken by terrestrial or airborne handheld small digital cameras by a rectification process to the façades of the buildings with the Photomodeller program. Obviously this 'beautification process' is useful for city centres or the most prominent buildings (public buildings, churches, railway stations).
- it is then possible to develop flythroughs through the cities for the creation of animations with the Erdas 'Virtual GIS' program.

Figures 4.25, 4.26 and 4.27, and Plates 22 and 23 in the colour section show the sequences of operation in obtaining a 3D city model:

- digital surface model (DSM) obtained by image correlation from photogrammetry.
- digitization of buildings from a vector GIS.
- DEM from image correlation subtracting building area signals.
- creation of building blocks after vector digitization of building outlines using height differences between DSM and DEM for the house areas.
- superposition of the digital orthophoto onto the ground.
- generation of roof structure from DSM height level differences for the buildings.
- pasting of texture information to building façades from handheld digital camera images.

Spatial analysis applications

There is a wide variety of spatial analysis applications, which can be demonstrated by the following examples. Figure 4.28 shows a demographic analysis for statistical data obtained for the counties of the USA.

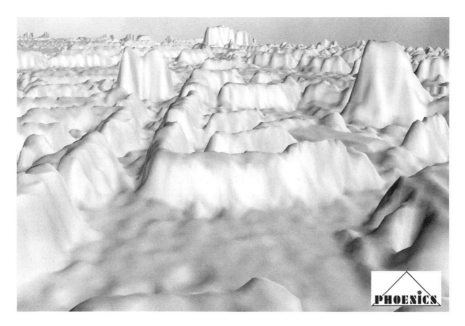

Figure 4.25 DSM via image correlation.

Source: Image courtesy of Phoenics GmbH, Hannover.

Figure 4.26 DEM via image correlation after subtraction of buildings.

Source: Image courtesy of Phoenics GmbH, Hannover.

Figure 4.27 Creation of buildings from different heights DEM–DSM.
Source: Image courtesy of Phoenics GmbH, Hannover.

Plate 24 in the colour section shows a municipal application for the different governmental, residential and commercial housing in the city of Bangkok.

Plate 25 in the colour section shows ownership of parcels in individually owned, business owned and institutionally owned land in a US city.

Plate 26 in the colour section shows the occurrence of a beetle infestation in New York City by geographic location.

Plate 27 in the colour section shows a road network with the analysed travel distance to the nearest fire station in a US city (black = less than 1.1 miles, to red 4.3 miles).

Plate 28 in the colour section shows the frequency of burglaries by blocks in a US city.

Plate 29 in the colour section shows the extent of industrial air pollution in a city (light blue = very low, dark blue = heavy).

Emergency GIS over crisis areas

The following illustrations show the compilation of a crisis area GIS of the Kosovo region of Serbia. Plate 30 in the colour section shows the layers of the crisis management GIS:

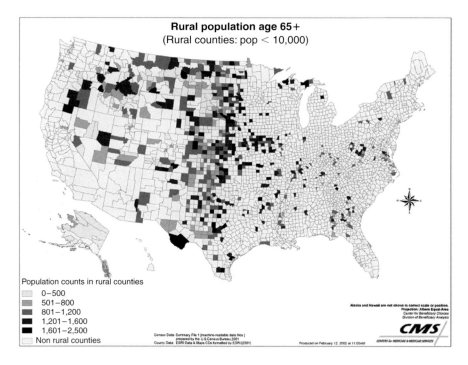

Figure 4.28 Rural population with age over 65 in the US counties.

Source: Map courtesy of Applied Geographics Inc., Boston, MA., and is used herein with permission by ESRI, Redlands, CA and Applied Geographics.

- a satellite image database.
- a topographic NATO map.
- additional GPS-related informations.
- a digital elevation model.
- the European CORINE land cover map.

Plate 31 in the colour section shows the ERS 1/2 radar interferometry generated DEM of the area.

Plate 32 in the colour section shows the satellite image database draped over the DEM.

Tourist information systems

A final example of tourist information systems is shown for a Tourist GIS made commercially available by the company Geospace. It displays, on the basis of a digital photomap, the:

- street names.
- house numbers.
- telephone booths.
- restaurants.
- hotels.
- tourist sites.
- pharmacies.
- hospitals, and other relevant information (see Figure 4.29).

Figure 4.29 Example of a tourist GIS for the city of Bonn, Germany.

Source: Image courtesy of Geospace GmbH, Köln.

5   Positioning systems

The capability to geocode GIS information in vector and raster form today depends on modern geopositioning systems.

The Global Positioning System (GPS)

The NAVSTAR–GPS (Navigation System with Time and Ranging – Global Positioning System) has been under development by the US military since 1973. From 1993, the system has consisted of twenty-one active satellites and three spares. The satellites orbit the earth at an altitude of 20 200 km. Their orbits are arranged in such a manner that at least four of them are in direct line of sight at any point of the earth's surface, 24 hours a day, unless the line of sight is obstructed by buildings, vegetation or steep topography. The use of GPS has been developed for real-time navigation and for geodetic positioning. Figure 5.1 illustrates the global GPS satellite configuration.

GPS signals

The satellites transmit signals in two carrier frequencies:

L1 at 1575.42 MHz, and
L2 at 1227.60 MHz.

Onto these carrier waves, specific codes are modulated which permit the measurement of distances from a particular satellite to a receiving antenna on a GPS receiver. Figure 5.2 shows the modulation of the carrier wave by codes.

For three known satellite positions on a common reference frame, the receiver coordinates can be computed on that reference frame by the three distances measured. A fourth satellite, however, needs to be observed due to the lack of synchronization between satellite clocks and the receiver clock. The ephemeris data of all twenty-four satellites are continuously determined by the operational network of control stations at Colorado Springs, Ascension, Diego Garcia, Kwajalein and Hawaii.

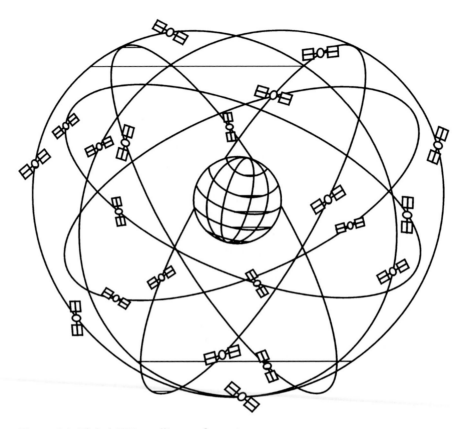

Figure 5.1 Global GPS satellite configuration.

Figure 5.3 shows the determination of the position from the four distance measurements, R_1, R_2, R_3 and R_4.

The measurement of distances between satellite and receiver is made possible by the codes modulated onto the carrier frequencies. There is a P-code (precision code) available to military users in encrypted form with a frequency of 10.23 MHz corresponding to a distance of 30 m. Civilian users must utilize the C/A code (clear/acquisition code) with a frequency of 1023 MHz corresponding to a distance of 300 m. The receiver has a copy of the code, which is shifted in time steps and correlated with the observed signal. The P-code permits direct distance observations with an accuracy of 0.6 m to 1 m. The C/A code permits the observation of distances with an accuracy of 6 m to 10 m.

Low-cost satellite receivers permit the direct utilization of the code observations. Higher precision and accuracy is achievable with high-cost satellite receivers, which permit the utilization of the phase of the received

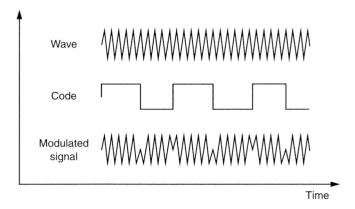

Figure 5.2 Modulation of carrier wave.

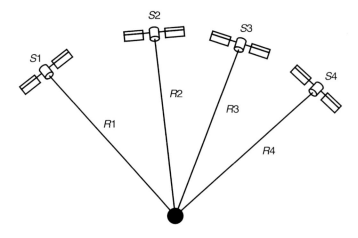

Figure 5.3 Absolute positioning.

carrier waves in addition to the C/A code. Depending on the observation time, the constellation of satellites and the atmospheric conditions carrier phase receivers are able to determine receiver positions to an accuracy or at least to a precision of 1 cm to 2–3 mm.

Code and carrier wave signals are receivable every second within a specific data block. Other blocks transmit data on the clock parameters, the broadcast ephemeris of the satellite observed and, at larger time spacings, the ephemeris data for all available GPS satellites plus ionospheric correction parameters. Up until mid-2000, the C/A code was artificially disturbed up to 100 m by the so-called 'selective availability' limiting the use of low-cost satellite receivers to that accuracy.

For this reason, differential GPS (DGPS) has been introduced as an operational observation technology. Even for carrier phase measurements, the DGPS technique has proved useful, since phase measurements are affected by an ambiguity term expressing the unknown number of complete wavelengths in the observed distance. It is deduced from the C/A code. But if it is faulty, a cycle slip by a full wavelength may occur.

Differential GPS

In DGPS, a master GPS receiver is utilized at a geodetic reference station. It receives the same satellite signals observed at a transportable second receiver (rover), which is used for the measurements at new point locations (see Figure 5.4).

In this way, relative positioning becomes possible with much higher accuracies. For distances between rover and master receivers of under 10 km, relative coordinate determinations in the range of ±1 cm in position and ±3 cm in ellipsoidal height become possible.

The disadvantage of using master–rover combinations for DGPS observations lie in the fact that two receivers have to be operated simultaneously. Nevertheless, this has proved feasible for airborne GPS observations for which accuracies of ±10 cm to 15 cm for the exposure station positions could be reached. For terrestrial measurements, there has been the tendency in a number of countries (e.g. Germany, Sweden, Austria, the Emirate of Dubai) to establish a number of permanently receiving master stations, which offer DGPS correction through radio transmission or through a mobile GSM network to the rover stations. The German SAPOS network,

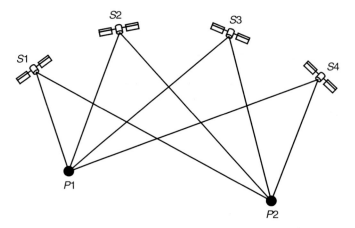

Figure 5.4 Relative positioning.

by the state survey administrations, provides such services with permanent receiver stations located at 50 km intervals.

The error sources for GPS signals stem from the following sources:

- ionospheric influences.
- tropospheric influences.
- multipath effects at the rover stations, which are station dependent.
- receiver antenna eccentricities, which can be eliminated through calibration.

Of these, the ionospheric influences are the most serious. Two German companies – Geo++ of Garbsen, near Hannover and Terrasat of Höhenkirchen, near Munich – have specialized in software systems to model the ionospheric effects from the observations of several permanent reference stations and to transmit the local ionospheric corrections to users, so that they can receive the corrections within less than a minute at the receiver with accuracies in the 1 cm range for position and 3 cm for ellipsoidal elevation within the 50 km spaced network of permanent reference stations in the real-time kinematic (RTK) mode of operations. This requires resolution of the ambiguities in static mode before the rover starts moving in phase lock.

On a much wider scale, a permanent reception station is available near Frankfurt (Main), which transmits DGPS corrections via long-wave radio transmission enabling central European users to reach real time positioning in the 1 m range.

Such international stations are interconnected worldwide to a civilian 'cooperative international GPS network' (CIGNET), which releases global ephemeris information covering an 8-day period through the US Coast Guard GPS information centre and has about twenty international tracking sites. The information may be utilized for post processing.

GPS satellite geometry and GPS software

The accuracy of GPS positioning depends on the accuracy of range measurement, σ_R, and the satellite configurations. The positioning error, σ_P, can be expressed as:

$$\sigma_P = \text{PDOP} \cdot \sigma_R.$$

$$\text{PDOP} = \frac{1}{V}$$

with V being the volume of the tetrahedron between the satellite positions and their vectors to the receiver.

In least squares adjustment of this geometric configuration, the covariance matrix C becomes:

$$C = \sigma_R^2 (A^T A)^{-1}$$

and

$$\sigma_P^2 = \sigma_R^2 (q_{xx} + q_{yy} + q_{zz})$$

Thus, PDOP corresponds to the square root of the trace of the covariance matrix.

Each receiver manufacturer (e.g. Trimble, Leica, Ashtech, Topcon) uses their own data format for calculations of positions, including accuracy indications such as the PDOP (GP Survey for Trimble, SKI for Leica, GPPS for Ashtech).

General purpose post-processing programs require a common translator program such as RINEX. These general purpose programs for multistation adjustments have been written by research institutes, e.g. BERNESE by the University of Berne in Switzerland, GEONAP by the University of Hannover in Germany or DIPOP by the University of New Brunswick in Canada. GEONAP uses the original phase data and BERNESE and DIPOP use their double differences.

GSM mobile phone location

GPS positioning depends on a direct line of sight between the satellite and the receiver; it therefore does not work in buildings or under dense vegetation. In areas where dense mobile telephone networks exist, it is possible to build up a location system which works in areas within reception reach. For this, the positions of the mobile phone transmission and reception antennas at which mobile phone calls are received must be known. These form the cells of the network. In urban areas, the design of these cells is on average about 700 m wide, in rural areas they can reach dimensions of tens of kilometres.

The identification of the cell is possible by use of slightly different frequencies. Within the cell a transmitted time code permits the measurement of the distance between cell antenna and the mobile phone. By intersection of at least two, three or more distances between the cell antennas and the mobile phone, a position is determinable. Tests for uses of emergency location systems in German cities have resulted in positioning accuracies of about 100 metres.

6   Cost considerations

At a time when various technical possibilities are available for the acquisition, management and updating of geoinformation, it is important to consider cost aspects. Scientists have generally been reluctant to make cost comparisons. They left suggestions to the forces of the market. But in the past, the geoinformation technologies followed slow traditional developments in the disciplines concerned. In the rapidly changing technology of today, cost considerations are important to judge which methodology is best utilized for a certain application.

There is an additional difficulty. When discussing costs, these are only rough indicators for actual prices in an open market. In general, a price for a product or for a service is composed of a cost figure plus an overhead plus the risk to get remunerated. Overheads vary greatly with salary structures of administrative services and the amount of taxes to be paid. Western Europe has considerably higher overhead percentages than Eastern Europe or countries in developing continents.

Price could be as much as 200 per cent of cost. If the services are provided on contract to a distant country, the risk increases and prices may be as much as 300 per cent of cost. Due to high labour costs in the developed world, automation-derived products are more favoured in the service-oriented high-tech economies. For these products, lower overhead costs may be charged.

In the age of a global economy, the mapping industry is in a transient stage. It is characterized by global partnerships on the one hand, and by an attempt to diversify the products to regional enterprises, where they are produced cheapest.

A few examples of cost strategies, based on world market prices, are given here.

Costs of aerial photography, orthophotography and topographic line mapping

Aerial photography costs consist of a base cost for mobilization of the aircraft plus a charge for the image. On average, the cost is $4000 plus $10 per image. The base cost may, of course, be considerably higher if the aircraft has to be transported to a distant region.

The orthophotography production with standard aerial cameras is based on costs per image:

- scanning   at $15/image
- aerial triangulation   at $25/image
- generation of a digital elevation model   at $120/image
- digital orthophoto generation   at $30/image
- mosaicking of digital orthophotos   at $20/image

The digital elevation model, whether automatically produced by image correlation or by measurement in stereo plotters, is the costliest part of the process. But it also generates the needed elevation base data.

If a digital elevation model is already available from other sources (previous topographic surveys, laser scanning or radar interferometry), then the orthophoto process is considerably cheaper. In any case, the orthophotography production is an automated process with little cost variation around the world.

Line mapping in stereo plotters or on screen on digital workstations, however, is a labour-intensive process. Depending on the topographic detail in an image pair or on a photograph, the restitution time can vary from 10 hours per model in rural areas to 100 hours per model in urban areas. For line mapping, labour prices of the developing countries are much more competitive than those of developed countries, provided that the quality of the product can be maintained. Labour costs per hour can vary, for example, from $25 per hour in India to $40 per hour in Germany.

The resolution of objects, the accuracy of their geometric measurement and the cost of the restitution are image-scale dependent. Figure 6.1 indicates the demand for panchromatic and multispectral resolution by various users.

Scale dependency versus cost is illustrated here with three examples.

An image scale of 1:6500 provides a resolution and a positional restitution accuracy of 6.5 cm. Scanned at 15 μm, it is suitable for the generation of 10 cm ground pixel orthophotos. This scale permits the mapping of utility manholes.

A photo covers an area of $a \times a = 1.5 \times 1.5$ km, and a neat model with a longitudinal overlap of 60 per cent and a lateral overlap of 30 per cent ($b = 0.4\,a$, $q = 0.7\,a$) covers an area of $0.28\,a^2 = 0.626$ km^2. This area can be flown at $10 plus proportional mobilization cost, and it can be produced into an orthophoto including DEM generation at $210. For an average restitution time of 50 hr per model, at $40 per hour, the line mapping cost for this area is $2000.

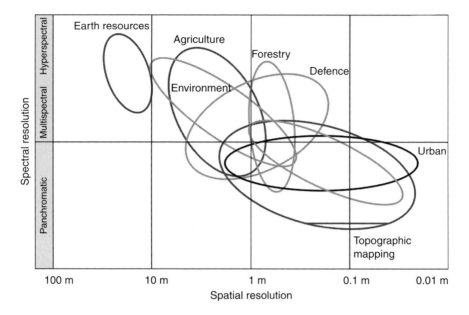

Figure 6.1 User requirements for resolution of imagery.

Source: Drawing courtesy of LH Systems (Leica Geosystems), San Diego, CA. © Leica Geosystems, 2002.

An urban area of 250 km² is covered by 400 photographs. The DEM has a standard deviation of ±10 cm. Therefore the project costs become:

aerial photography	$4000 + $4000 = $8000
scanning	$6000
aerial triangulation	$10 000
digital elevation model	$48 000
digital orthophoto generation	$12 000
mosaicking	$8000
total orthophoto cost	$92 000 or $368/km²
line mapping 1:1000	$800 000
total line mapping cost	$892 000 or $3568/km²

If the same area is covered at an image scale of 1:13 000, the resolution and the positional accuracy is 13 cm. A 20 cm ground pixel orthophoto may be generated. This scale is suitable for the mapping of buildings and road features. A photo covers an area of 3 × 3 km. The neat model area is 2503 km². The urban area of 250 km² is imaged by 100 photos. The DEM has a standard deviation of ±20 cm.

The project costs become:

aerial photography	$4000 + $1000 = $5000
scanning	$1500
aerial triangulation	$2500
digital elevation model	$12 000
digital orthophoto generation	$3000
mosaicking	$2000
total orthophoto cost	$26 000 or $104/km^2
line mapping 1:2000	$200 000
total line mapping cost	$226 000 or $904/km^2

From an image scale 1:40 000, a resolution and a positional accuracy of 40 cm can be achieved. Scanned at 12.5 μm, it generates 50 cm ground pixel orthophotos. This scale satisfies planning needs. A photo covers an area of 9.2×9.2 km^2. The neat model area is 23.7 km^2. The urban area of 250 km^2 is covered by eleven photos. The DEM has a standard deviation of ±60 cm.

The project costs become:

aerial photography	$4000 + $110 = $4110
scanning	$165
aerial triangulation	$275
digital elevation model	$1320
digital orthophoto generation	$330
mosaicking	$220
total orthophoto cost	$6420 or $25.68/km^2
line mapping 1:10 000	$22 000
total line mapping cost	$28 420 or $88/km^2

The example clearly demonstrates that the orthophoto is a much cheaper line map substitute. Line maps have their advantages for topological analysis in a GIS, but not all features need to be vectorized. Because of the cost element, some non-European countries prefer to restrict their vector database to the property cadastre. All relevant attributes can be attached to the parcel. Due to the importance and the sensitivity of land transactions, the cadastre including the attributes can be maintained on a real-time basis for each transaction. The cadastral vector database may be overlaid with the orthophoto in raster form to depict buildings and road features. If desired, buildings may be numbered and identified by pointers to attach attributes, without the need to show the outlines in vector graphics.

The utility network is essentially a line network between manholes, which can easily and rapidly be measured by RTK-GPS surveys as an additional vector layer for each utility type. Like the property cadastre, it can also be superimposed with the orthophoto. In surveying the utility manholes, their attribute information may be verified or added on-site. As the number of utility points for each utility is rather small, ground surveys are more effective and less costly.

GPS-supported ground surveys

Ground surveys are generally carried out for cadastral purposes of larger areas in terrain, where visibility requirements of the boundaries prohibit the use of photogrammetric techniques. They may also be preferable in countries where photo-adjudication is not accepted due to accuracy concerns.

In Europe, cadastral data are already existent. Therefore, a survey cost comparison with photogrammetric methods is not useful there. However, in a number of development projects, ground survey costs per parcel, including the land registration or land titling aspects, have been established.

In Albania, a new cadastre has been established by European funding at a cost of $5 per parcel. The procedure used was aerial photography–aerial triangulation–digital elevation model–digital orthophoto generation followed by public photo adjudication process.

In Georgia, a large German technical cooperation project was carried out by GPS-supported electronic tacheometers at a cost of $10 per parcel. A survey crew is able to measure about fifty parcels per day. In doing so, it has been proved useful to support the ground surveys with aerial photos or orthophotos. For this purpose, 'digital plane tables' in the form of large-screen PDAs may be used which record the measured GPS or electronic tacheometer measurements on the screen. These data are superimposed with preprepared (ortho-)photographic data on the screen, which helps to identify points to be measured terrestrially. A 'digital plane table' costs about $10 000.

An urban area of $250 \, km^2$ has about 80 000 parcels. The cost of surveying these terrestrially would therefore be $800 000, which is about the same as the photogrammetric line mapping cost at the scale 1:1000, but about four times as much as the photogrammetric mapping cost at the scale 1:2000. This corresponds to a terrestrial survey cost of $3200/km².

Digital elevation models

For the acquisition of digital elevation models, there exist a great variety of technical possibilities, as shown in Table 6.1.

Aerial triangulation versus direct sensor orientation

Another issue is the possibility of replacing the photogrammetric aerial triangulation process with the possibility of measuring sensor positions and sensors directly by GPS/IMU systems. The combined use of aerial triangulation and

Table 6.1 Comparison of acquisition costs for digital elevation models

Methodology	Height accuracy	Cost	Application
Satellite radar interferometry	±12 m	$2/km²	topography
Airborne radar	±2 m	$20/km²	topography
Aerial photography 1:40 000	±0.6 m	$25/km²	topography, city models
Aerial photography 1:13 000	±0.20 m	$100/km²	non-forested areas
Aerial photography 1:6500	±0.10 m	$350/km²	non-forested areas
Airborne laser	±0.15 m	$500/km²	forest, building heights, coastal areas
Ground surveys	±0.10 m	$1000/km²	dense forest

GPS/IMU observations in a bundle block adjustment is, in any case, advantageous for the limitation of the needed ground control points and for the increased quality control possibilities for improved reliability of geocoding.

But the acquisition of additional inertial sensor data significantly increases the investment costs for aerial photography. The same is true in the case where inertial sensor-dependent digital airborne scanners are intended to be used instead of standard aerial photography.

An airborne inertial sensor system will require acquisition of a $0.5 million investment. An inertial measurement unit by Applanix or IGI will cost between $0.5 million and $0.25 million as an investment. It needs to be balanced against the savings obtained by skipping aerial triangulation. This depends on the volume of work anticipated in order to justify these investment costs.

Mapping from space

Ever since Landsat TM initiated the provision of satellite imagery at 30 m ground pixels in 1982, there have been attempts to commercialize the market for space imagery. The US Landsat commercialization act of 1984 paved the way for the marketing of space images, particularly for 10 m resolution panchromatic Spot images by Spot-Image. While Landsat images covering 185 km × 185 km areas at 30 m pixels and Spot images covering 60 km × 60 km areas at 10 m in black and white and 20 m in colour, would cost several thousand dollars each, the prices for image products have been considerably lowered since. The recent lowering of Landsat prices to a few hundred dollars per scene, following the public domain principles of the US government, has brought a disturbance in the imagery market of medium-resolution satellites.

With the advent of high-resolution satellites by Ikonos 2 in 1999, with

Table 6.2 Carterra-Ikonos 2 products by space imaging

Type of data	Geometric accuracy	Price per km² (Nov. 2001)
Raw data	±12 m	$29/km²
Orbit corrected data	±6 m	$50/km²
Geocorrected data on the basis of ground control	±3 m	$100/km²
Stereo imagery	±1 m	$200/km²

1 m pixels in black and white and 4 m in colour, the Space Imaging company now provides a number of products listed in Table 6.2.

The Ikonos 2 products now compete with Russian space photography products of the Spin 2 program (KVR 1000 panoramic photographs digitized to 2 m pixels) with a geometric accuracy of ±5 m at $25/km².

Another advantageous alternative is the use of standard aerial photography at the scale of 1:40 000 with a much higher resolution of 0.5 m pixels at about $25/km².

7 Technological changes

A more philosophical approach to technological changes has been initiated by the studies of Oswald Spengler of Munich (1880–1936). His treatise on 'Cultural cycles of civilizations', in which he described the cycles in the development of civilizations – blossoming, maturity and decay – inspired the British historian Arnold Joseph Toynbee (1889–1975) to write his 'A study of history' in which he described the cycles of civilizations from their early beginnings to the present. The motivations for changes were *challenge* and *response*. Toynbee's ideas affected the studies of Pitirim Alexandrovich Sorokin (1889–1968), a Harvard professor on 'social and cultural dynamics'.

A technological link to these cycles was established by Nikolai Dmitrijevich Kondratjev (1892–1930), the founder of the Conjuncture Economics Institute in Moscow, the originator of the five-year plans for Soviet agriculture. His treatise on 'the major economic cycles' of 1926 postulates that economic changes are generated by new technology. In Figure 7.1, these economic cycles at about 50-year periods are generated by technological inventions. This initiates a rapid growth of the technology and its application.

It is of interest that, for example, the four stages of the development of the geoinformation discipline photogrammetry:

- single image photogrammetry, from 1850 to 1900
- analogue stereo photogrammetry from about 1900 to 1950
- analytical photogrammetry from about 1950 to 2000
- digital photogrammetry for the twenty-first century

also fall into this pattern.

The geoinformation disciplines of the twentieth century have favoured an ever-increasing specialization. In the twenty-first century, a renaissance of thought is in order, in which the various specializations must be considered together and used jointly for a sustainable development.

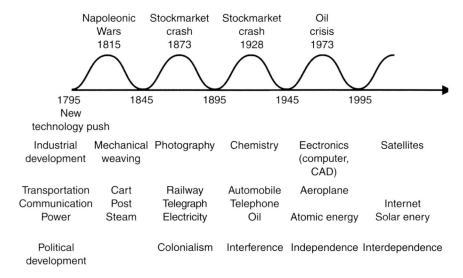

Figure 7.1 Kondratjev's economic cycles.

Bibliography

1 Introduction

Capters, F. and Decleir, H. *The World in Perspective: A Directory of World Map Projections*, John Wiley, New York, 1990.

Everest, G. *An Account of the Measurement of an Arc of the Meridian Between the Parallels 18°3' and 24°7'*, London, 1830.

Helmert, F.R. *Die mathematischen und physikalischen Theorien der höheren Geodäsie*, 1880, reprint Minerva, Frankfurt a.M., 1961.

Konecny, G. 'Paradigm Changes in ISPRS from the First to the Eighteenth Congress in Vienna', *ISPRS Highlights*, 1, 1, 1996, 10–15.

Longley, P.A., Goodchild, M.F., Maguire, D.J. and Rhind, D.W. *Geographic Information Systems*, 1 & 2, John Wiley, New York, 1999.

Seeber, G. *Satellite Geodesy*, De Gruyter, Berlin, 1993.

Smith, J.R. *Introduction to Geodesy: The History and Concepts of Modern Geodesy*, Wiley, New York. Series in Surveying and Boundary Control.

Torge, W. *Geodesy*, 3rd edition, De Gruyter, Berlin, 2001.

UN Secretariat. 'The Status of World Mapping 1990', *World Cartography*, 1993, UN Cartographic Conference, Beijing, Paper prepared by A. Brandenberger and S. Ghosh.

United Nations Statistic Division. *World Statistics Pocketbook*, Series IV, No. 21, UN Publications, New York.

2 Remote sensing

American Society of Photogrammetry and Remote Sensing. *Manual of Remote Sensing*, American Society of Photogrammetry, Falls Church, Va., 1975.

Askne, J. (ed.) *Sensors and Environmental Applications of Remote Sensing*, Proceedings of the 14th EARSeL Symposium, Göteburg, Sweden, Ashgate Publishing Company and A.A. Balkema International Publishers, Rotterdam and Brookfield, VT, 1995.

Campbell, J.B. *Introduction to Remote Sensing*, Taylor & Francis, London, 1996.

Canada Centre for Remote Sensing. *Fundamentals of Remote Sensing*, Tutorial, http://www.cors.nrcan.gc.ca/ccrs/eduref/tutorial/tutore.html.

Canty, M.J. *Fernerkundung mit neuronalen Netzen*, Expert Verlag Remingen-Matensheim, Germany, 1999.

CNES. 'Resources in Earth Observation', *Case Studies, Data and Information for*

Education and Developing Countries, 1st edition, CSIRO and Smith System Engineering Ltd, http://sirius-ci.cst.cnes.fr:8100/CD-ROM-97/astart.letm.

Colwell, R.N. *et al.* 'Basic Matter and Energy Relationships Involved in Remote Reconnaissance', *Photogrammetric Engineering*, periodical of the American Society for Photogrammetry, Falls Church, VA, 1963, 761–799.

Cracknell, J.P. and Hayes, L.W.B. *Introduction to Remote Sensing*, Taylor & Francis, London, 1991.

Gibson, P. and Power, C. *Introductory Remote Sensing: Principles and Concepts*, Taylor & Francis, London, 2000.

Hapke, B. *Theory of Reflectance and Emittance Spectroscopy*, Cambridge University Press, Cambridge, 1993.

Henderson, F.M. and Lewis, A.J. 'Principles and Applications of Imaging Radar', *Manual of Remote Sensing*, 3rd edition, vol. 2, John Wiley, New York, 1998.

Lillesand, T.M. and Kiefer, R.W. *Remote Sensing and Image Interpretation*, John Wiley, New York, 1979.

Lippmann, R.P. 'An Introduction to Computing with Neural Nets', *IEEE ASSP Magazine*, April 1987, 4–22.

Milman, A.S. *Mathematical Principles of Remote Sensing*, Sleeping Bear Press. Chelsea, MI.

Nagler, T. and Rott, H. *Overview of Current and Planned Spaceborne Earth Observation Systems*, Space Application Institute, Joint Research Centre, Ispra, 1998, EUR 18673 EN.

Pelzer, H., Crampéa, F. and Rosen, P. 'The Mir 7.1, Hector Mine, California Earthquake: Surface Rupture, Surface Displacement Field and Fault Ship Solution from ERS SAR Data', *Earth and Planetary Sciences*, 333, 2001, 545–555.

Rencz, A.N. and Ryerson, R.A. (eds). *Manual of Remote Sensing*, vol. 3, *Remote Sensing for the Earth Sciences*, 3rd edition, John Wiley, New York, 1999.

Richards, J.A. and Xinping Jia. *Remote Sensing Digital Image Analysis*, 3rd edition, Springer, Berlin, 1998.

Sabins, F.F. *Remote Sensing – Principles and Interpretation*, 3rd edition, W.H. Freeman, New York, 1996.

Schanda, E. *Physical Fundamentals of Remote Sensing*, Springer Verlag, Berlin and Heidelberg, 1986.

Schowenderdt, R.A. *Remote Sensing – Models and Methods for Image Processing*, 2nd edition, Academic Press, San Diego, CA, 1997.

Short, N.M. Sr. *The Remote Sensing Tutorial – an Online Handbook at NASA's Goddard Space Flight Center*, http://rst.gsfc.nasa.gov.

SPIE. *Proceedings of the Conference on Commercial Remote Sensing Platforms and Applications*, International Society for Optical Engineering, Bellingham, WA, 1999.

Spiteri, A. (ed.) *Remote Sensing – Integrated Applications for Risk Assessment and Disaster Prevention for the Mediterranean*, Proceedings of the 16th EARSeL Symposium, Ashgate Publishing Co. and A.A. Balkema, Rotterdam, Brookfield, VT 1997.

Steinborn, W. and Sprengelmeier-Schnock (eds) *Raumfahrt zum Nutzen Europas – Die Perspektiven der Fernerkundung mit Satelliten*, Herbert Wichmann Verlag, Karlsruhe & Heidelberg, 1993.

3 Photogrammetry

Ackermann, F. 'Airborne Laser Scanning – Present Status and Future Expectations', *ISPRS Journal of Photogrammetry and Remote Sensing*, 54, 2–3, 1999.

Ackermann, F., Ebner, H. and Klein, H. 'Ein Rechenprogramm für die Streifentriangulation mit unabhängigen Modellen', *Bildmessung und Luftbildwesen*, 1970, 206–217.

Baltsavias, E.P. 'Airborne Laser Scanning: Basic Relations and Formulas', *ISPRS Journal of Photogrammetry and Remote Sensing*, 54, 2–3, 199–214.

Ebner, H. 'Zwei neue Interpolationsverfahren und Beispiele für ihre Anwendung', *Bildmessung und Luftbildwesen*, 1979, 15–27.

Fricker, P. 'ADS 40 – Progress in Digital Aerial Data Collection', *Photogrammetric Week*, 2001, 105–116.

Greve, C. (ed.). 'Digital Photogrammetry', an addendum to the *Manual of Photogrammetry*, American Society of Photogrammetry and Remote Sensing, Bethesda, MD, 1996.

Heipke, C., Jacobsen, K. and Wegmann, H. *The OEEPE-Test on Integrated Sensor Orientation – Analysis and Results*, OEEPE Workshop Integrated Sensor Orientation, Hannover, Sept. 2001.

Hinz, A., Dörstel, C. and Heier, H. 'DMC – The Digital Sensor Technology of Z/I Imaging', *Photogrammetric Week*, 2001, 93–104.

Hobrough, G.L. 'Automatic Stereo Plotting', *Photogrammetric Engineering*, 1959, 763–769.

Konecny, G. and Lehmann, G. *Photogrammetry*, De Gruyter, Berlin, 1984.

Konecny, G., Kruck, E. and Lohmann, P. 'Ein universeller Ansatz für die Geometrische Auswertung von CCD-Zeilenabtasteraufnahmen', *Bildmessung und Luftbildwesen*, 54, 1986, 139–146.

Kraus, K. 'Interpolation nach kleinsten Quadraten in der Photogrammetrie', *Bildmessung und Luftbildwesen*, Journal of the German Society of Photogrammetry, Herbert Wichmann Verlag, Karlsruhe, 1972, 7–12.

Kraus, K. *Photogrammetry*, 4th edition, Dümmler Verlag, Stamm GmbH, Köln, 1994.

Kremer, J. 'CCNS and Aerocontrol: Products for Efficient Photogrammetric Data Collection', *Photogrammetric Week*, 2001, 85–92.

Luhmann, T. *Nahbereichsphotogrammetrie*, Herbert Wichmann, Heidelberg, 2000.

Mikhail, E., Bethel, J.S. and McGlone, C. *Introduction to Modern Photogrammetry*, John Wiley, New York, 2001.

Mostafa, M., Hutton, J. and Raid, B. 'GPS/IMU Products – The Applanix Approach', *Photogrammetric Week*, 2001, 63–84.

Petrie, G. '3D Stereo Viewing of Digital Imagery', *Geoinformatics*, 4, 2001, 24–29.

Read, R. and Graham, R. *Manual of Aerial Survey: Primary Data Acquisition*, Whittles Publishing, Caithness, 2001.

Schenk, T. *Digital Photogrammetry*, vol. 1, Terra Science, Lamelville, OH, 1999.

Schroeder, M. *Zur photographischen Aufnahmetechnik im Weltraum für kartographische Aufgaben*, Wiss. Arbeiten der Fachrichtung Vermessungswesen, Universität Hannover, Nr. 165.

Scott, C. *Introduction to Optics and Optical Imaging*, John Wiley, New York, 1997.

Slama, C.C. (ed.) 'American Society of Photogrammetry and Remote Sensing', *Manual of Photogrammetry*, 4th edition, Falls Church, VA, 1980.

4 Geographic information systems

AMD. *The AMD x86-64 TM Architecture*, Programmers overview, Publication 24108, Jan. 2001.

ATKIS. *Amtliches Topographisch-kartographisches Informationssystsem*, http://www.atkis.de.

Bartelme, N. *Geoinformatik*, 3rd edition, Springer, Berlin–Heidelberg–New York, 2000.

Berry, J.K. *Beyond Mapping: Concepts, Algorithms and Issues in GIS*, John Wiley, New York, 1996.

Bill, R. *Grundlagen der Geoinformationssysteme*, vol. 2, Wichmann, Heidelberg, 1991.

Bill, R. and Fritsch, D. *Grundlagen der Geo-Informationssysteme*, vol. 1, Wichmann, Karlsruhe, 1991.

Chou, Yue-Hong. *Exploring Spatial Analysis in GIS*, OnWord Press and Delmar, Thomson Learning, Albany, NY, 1996.

de By, R.A. *et al*. *Principles of Geographic Information Systems – an Introductory Textbook*, International Institute of Aerospace Survey and Earth Sciences (ITC), Enschede, 2001.

de Mers, M.N. *Fundamentals of Geographic Information Systems*, John Wiley, New York, 1999.

ESRI. *Map Objects Internet Map Server Reference* (issued with software), ESRI, Redlands, CI, 1998.

FIG. *Reforming and Benchmarking the Cadastre: Measuring the Success*, Proceedings of the FIG Comm 7 Symposium in Gävle, June 2001.

Gifford, F. 'Internet GIS Architecture – Which Side is Right for You?', *GeoWorld*, 1999.

Heipke, C. *GIS Course Rawalpindi 1999* (Components of GIS Hardware, Components of GIS Software, Introduction to GIS Data Bases), CD-ROM Institute for Photogrammetry and GeoInformation, University of Hannover.

Kaufmann, J. and Stendler, D. *Cadastre 2014 – A Vision for a Future Cadastral System*, FIG 1998, http://www.swisstopo.ch/fig-weg71/cad 2014/cad2014/index.htm.

Korn, G.A. and Korn, T.M. *Mathematical Handbook for Scientists and Engineers*, McGraw Hill, New York–Toronto–London, 1961.

Korte, G. *The GIS Book*, 5th edition, OnWord Press and Delmar, Thomson Learning, Albany, NY, 2000.

Linder, W. *Geo-Informationssysteme*, Springer, Berlin, 1999.

Longley, P. and Batty, M. *Spatial Analysis: Modelling in a GIS Environment*, John Wiley, New York, 1997.

Marshall, J. 'Developing Internet-Based GIS Applications', http://www.giscafe.com/TechPapers/Papers/paper 058/p405.htm.

Mitchell, A. *The ESRI-Guide to GIS-Analysis, Vol. 1: Geographic Patterns and Relationships*, ESRI Press, Redlands, CA, 1999.

Murai, S. *Textbook on Remote Sensing and GIS*, 1999: 'Remote Sensing Notes' produced by the National Space Development Agency of Japan (NASDA); 'GIS Workbook' produced by the Asian Center for Research and Remote Sensing (ACRoRS) at Asian Institute of Technology (AIT).

National Spatial Data Infrastructure: *The 2002 National Spatial Data Infrastructure Cooperative Agreements Programme (CAP)*, http://www.fgdc.gov.

Standish, T.A. *Data Structure Techniques*, Addison-Wesley, Reading, CA, 1980.
Sylla, C.I., Wade, I.A., Hengue, P. and Gerbe, E. 'Environmental Information Systems in Sub-Saharan Africa', Country Case: Senegal, Report published by the GTL, Eschborn, Germany, 1997.
Winget Godin, L. *GIS in Telecommunications Management*, ESRI Press, Redlands, CA, 2000.
Zeller, M. *Modelling Our World*, ESRI Press, Redlands, CA, 1999.

5 Positioning systems

Kaula, W.M. *Theory of Satellite Geodesy*, Dover Publications, Mineda, NY, 2000.
Seeber, G. *Satellite Geodesy: Foundations, Methods and Applications*, De Gruyter, Berlin, 2000.
Teunissen, P.J.G. and Kleusberg, A. (eds) *GPS for Geodesy*, 2nd edition, Springer, Berlin–Heidelberg–New York, 1998.

6 Cost considerations

Betz, R. and Schott, B. *Mehr Umsatz mit Perfect-Marketing*, Max Schirm Verlag, Germany, 1996.

Index